EMPLOYABILITY OF FRESH ENGINEERS

Issues and Challenges

Edited by

Dr. Sabu Thomas

Prints Publications Pvt Ltd

New Delhi

Published by

Prints Publications Pvt Ltd
Viraj Tower-2, 4259/3, Ansari Road,
Darya Ganj, New Delhi-110002
Tel. : +91-11-45355555
Fax: +91-11-23275542
E-mail : contact@printspublications.com
Website : www.printspublications.com

First Edition : 2022 (Hardbound)

ISBN: 978-93-936740-9-8

Price: ₹ 1295/-

Published and Printed by Mr. Pranav Gupta (Director) on behalf of Prints Publications Pvt Ltd, New Delhi.

Foreword

I am happy to know that the proceedings of the National Seminar on 'Employability of Fresh Engineering Graduates in the Era of Globalisation: Problems and Prospects', conducted by the Nodal Centre for Kerala of National Technical Manpower Information System (NTMIS) is being published. The subject matter is deliberated in depth and covers areas like pre-eminence of technical education, the requirement of preserving 'Quality over Quantity' in the area of technical education, the imperatives of academia industry collaboration for their mutual survival and sustained growth. The papers that are included in the book are of significant contemporary relevance and are worthy of a studied discussion for the building up of the nation and economy.

Publication of the proceedings will serve the needs of end users such as policy makers, education providers, students and the parents. The Co-ordinator deserves special appreciation for having organized such a worthy discourse that helps to provide strong impetus for the preservation of quality in Technical Education.

March 2013

Dr. RamachandranThekkedath
Vice Chancellor
Cochin University of Science and Technology

Preface

Technical education is the backbone of the labour market which functions as a catalyst for the economic development of any country. The prospects of technical development rest among others on the timely education in the different technical fields suited to the State. For this technical education has to be planned in such a way that there be a proper balance between general education and technical education at all levels. The actual intake and outturn in all technical fields constitutes a major part of labour force in the state. A study on the growth of technical education at institutions and the rate of intake and outturn of technically trained people who enter in to the labour market and the employability of fresh engineering graduates and diploma holders and the problems they face in the era of globalisation is an issue that needs to be discussed at all levels. To discuss these issues, a two days' seminar was conducted at CUSAT, Kochi. Leading technologists, academicians, industrialists, policy makers and researchers have deliberated on various issues related to the employability of fresh engineering degree and diploma holders in the Indian manufacturing and services sector.

The seminar was organised by NODAL CENTRE FOR KERALA of National Technical Manpower Information System, Cochin University of Science and Technology, Kochi, Kerala in association with the Kerala State Higher Education Council (KSHEC), National Productivity Council (NPC), New Delhi, Institute of Applied Manpower Research (IAMR), Delhi and with the financial support by Indian Council of Social Science Research (ICSSR), New Delhi, Tourism Department, Government of Kerala. ***Publication of the edited volume of the proceedings of the seminar is with the active support of*** *the National Productivity Council, New Delhi*. As the Co-ordinator, I express my sincere thanks to the Hon'ble Vice Chancellor, CUSAT and all the institutions associated with the Seminar. My thanks are also due to the paper presenters and participants who attended the programme. Professor. Joseph Alexander, the Project Officer and Head of the Nodal Centre for Kerala had provided all the support for the event to be a grant success. I express my sincere thanks to him. Finally, all the staff members in the Nodal Centre, who have provided whole hearted support, deserve sincere thanks.

I am very grateful to Mr. Pranav Gupta, Director of Prints Publication Pvt. Ltd, New Delhi, for his cooperation and support for bringing out this book.

Sabu Thomas

Contents

Seminar Papers

on

EMPLOYABILITY OF FRESH ENGINEERS

1

Employability of Fresh Engineering Graduates in the Era of Globalisation

Problems and Prospects

*Sabu Thomas

Introduction

Employability of fresh engineering graduates continues to be a major problem faced by the technical education sector in India. Due to the lack of core competencies among the fresh engineering pass outs, the availability of technical manpower having advanced technical knowledge and skills for the industries in the country are compromised. It has been widely acknowledged in recent years that technical education in India has failed to keep pace with the requirements of the industry and kept on churning out engineering graduates and diploma holders who are generally found

* Dr. Sabu Thomas, Co-ordinator, National Seminar on "Employability of Fresh Engineering Graduates in the Era of Globalisation: Problems and Prospects" is presently the Co-ordinator, K M Mani Centre for Budget Studies(KMMCBS), Cochin University of Science and Technology (CUSAT), Kochi -682 022

unemployable by the manufacturing sector. Recent mushrooming of technical institutions in both the private and public domains focused mainly on quantity rather than quality of technical education. Many fresh engineering pass outs are reported to be lacking in much needed technical capabilities required in the manufacturing sectors. While Indian industries face shortage of skilled technical manpower, there is increasing unemployment among the fresh engineering graduates and diploma holders. This scenario is forcing many technically qualified fresh engineering graduates to move to non-technical areas in search of a livelihood, hence, the nation is losing precious technical manpower after spending a lot of resources

The book comprises of 15 papers presented during the two day national seminar. The papers throw light to the problems facing the fresh engineering graduates in the country.

In the paper on 'Quality of Education in the newly formed Engineering Colleges and Polytechnics – Issues and Challenges', Shri. B.Anil has presented the issues and challenges faced by the technical education sector in India in the wake of the mushrooming of new generation institutions taking the example of the scenario in Kerala. He stressed the need for Quality Assurance (QA) strategy as a possible way to solve the quality issues of technical educational institutions.

In the paper on 'Role of Modular Courses/ Finishing Schools in filling the gap between Education and Employability' Smt. Shakila Shamsu noticed that currently India is passing through an unprecedented phase of demographic change which will result in the increase of the size of labour force in the country. She has also suggested certain strategies for enhancing the range of employability factor of our students through varied range of modular and skill based courses.

In her second paper, Smt. Shakila Shamsu, which is on 'Industry Academic Linkages in the Global Era- Issues and Employability of Engineers' stresses the growing need and significance for the collaboration between academic institutions and industry in the country's economic scenario.

Industry Academic Linkages and Employability of Fresh Engineers – Review of Recent Trends in India: In the paper presented jointly by Dr.S.K.Chakraborthy and Dr. K.P Sunny, the authors explore the issues relating to the importance of Industry- Academic Linkages and the need for setting up finishing schools or modular courses in the engineering and polytechnic colleges with the active participation of Industry. They have highlighted Government's national manufacturing policy which promises employment generation to the tune of 100 million persons in the manufacturing sector by 2022. For this the technical manpower skills have to be effectively imparted in their respective trades for the development of the manufacturing sector in India.

A paper presented jointly by Prof. U. Upadhyaya and Vandita Sahay on 'Problems of Quality of Engineering Graduates, Skill Development and Related Issues' the authors discuss on the market demand for young engineers in the country in terms of quality recession largely unmet because most graduates who come for job interviews fall short of required quality standards due to lack of quality orientation of engineering education in the country.

The paper on 'Labour Market Situation of Agricultural Manpower: Some Issues' I.C Awasthi and Indra Kumar have discussed the concern of agricultural sector, educational challenges in the perspective of new agriculture, the institutional capacity, intake capacity, outturn pattern and shortage of teachers and the labour market features in the agriculture sector. The paper pointed out the dramatic fall of outturn of agriculture engineers and gross mis-utilisation of human resources in terms and outturn of agriculture engineers. It has also mentioned that almost one third of agriculture graduates are unemployed. Emoluments of agriculture engineering graduates are much lower compared to other disciplines.

Dr. M.P. Chandra Sekharan, the presenter of the paper on 'Employability of Engineering Graduates' summarised that ability without knowledge and skill is hollow in the case of an engineering graduate.

Failure rates and their determinants in the engineering degree courses in Kerala: Factor Analysis of a socially deprived group: The paper prepared by Dr.K. Pushpagadhan analyzed the nature of causes of failure rates among students in engineering degree courses in Kerala for the period 1991-2011. The pass rates of students from socially deprived classes are much lower than the students from socially advanced classes. The determinants of failure rates among most socially deprived classes of students as stated by Dr.

Pushpangadan are; i. the quality of training in Mathematics, Physics and English at the secondary and higher secondary schools, ii. Family support, iii. teachers support, iv. place of stay, v. learning atmosphere, and vi. learning practices.

Joint paper by Mr. Ramesh U., Biji. C.V. and Manoj R. on 'Relevance of English Language Teaching in Technical Education Curriculum to face the challenges of Globalisation' has emphatically stated the need and importance of teaching English Language in the technical institutions and further stated that even students who have studied all along in English medium schools find it hard to communicate effectively in English. They have concluded the paper stating that, no matter how intelligent and knowledgeable a person is, no matter how many degrees he may possess, he shall not be successful in getting into an appropriate career until and unless he possesses excellent and effective communication and listening skills in English language.

The paper on 'Self financing colleges and Employability of Engineering Graduates': by Dr. Usha Menon, stated that our engineering graduates should have the know how to contribute positively to the industrial and business development of our country. Hence the real challenge before the teaching community is to function as neutral ambassadors in providing inputs that are required through scientifically designed curriculum and assessment to meet the demand of the industry today.

Sri. K.V.Varkey in his paper on 'Emerging Job Opportunities for Fresh Engineers' briefly mentioned some new developments in business/ technicality, which may result in the death of some of the existing job opportunities as well as generation of new employment avenues for fresh engineering graduates and diploma holders. He pointed out that the employers in collaboration with university have to make it a practice to impart continuing education for their engineers in latest technologies, so that their services can be utilized more effectively for a long period.

The paper on 'Profile of Technical Education Facility and Employment Scenario in Karnataka State' focuses on some of the vital parameters related to technical education such as growth of institutions, sanctioned intake, actual admissions for the year 2005 to 2009 in Karnataka and teacher student ratio, employment profile such as first paid activity and their salary for the degree and diploma engineering for the Karnataka state during the batch year 2009. This paper presented by Dr. Dodamani and Arun Kumar Shetty. B concluded that the percentage of paid employment was encouraging among degree holders for the year 2009.

Academic excellence and improvement of employability of student outputs should be among the top priorities of engineering colleges to ensure a good supply of labour to satisfy the industry demand of our burgeoning economy as stated in the paper on 'Engineering Labour Market in Kerala - Quantity and Quality Issues: presented jointly by Miss. Dhanya Menon and Zakkaria K.A.

The engineering graduates of the advanced countries possess a high level of confidence in handling unknown problems, but the engineering graduates in our country are deficient in handling unknown problems as stressed in the paper on 'Employability of fresh Engineering Graduates: Problems and Prospects' presented by Mr. Murali Pasupathy in the seminar. Hence in addition to unemployment, we are bringing in a kind of under employment for engineers where people do work that is much below the skill- level, that they are supposed to have.

Sri. M.K Shelly in his paper on 'Employment Scenario of Fresh Degree and Diploma Holders: Problems of Unemployment among the Engineers in Kerala' has stated that, the employment of skilled engineers and diploma holders is a major problem despite the fact that the number of engineering colleges are increasing on a day -to-day basis. In the paper he has highlighted the employment status of the fresh engineers passed out in the year 2009 from Kerala. The paper describes various dimensions of the problems of unemployment. It is also concluded that the problem of unemployment among engineers is acute for both the diploma holders and graduate engineers. It is also observed that the gender characteristics of the students also influence the prospects of employment and the incidence of unemployment is found more among female engineers in the state.

Five other papers have also been presented in the seminar, but due to non-availability of final paper from the respective presenters, those papers could not be included in this book.

2

Quality of Education in the Newly formed Engineering Colleges and Polytechnics

Issues and Challenges

*B. Anil

ABSTRACT

The economic development of a country is strongly linked to education – more importantly to the technical education provided. Engineering colleges and Polytechnics can be viewed as critical institutions of a nation which produce a corps of manpower that serve the community and the entire nation in its nation building exercise. They also serve as a means of social advancement and economic empowerment to the beneficiary (student). In India, technical education is imparted at various levels: craftsmanship, diploma, graduate and post-graduate level and research in specialised fields, catering to various aspects of technological development and economic progress. In order to sustain the growth rates of the economy the pool of

* Principal, Govt. Engineering College Wayanad, Kerala, (Chairman, ISTE Kerala section)

technical talent needs to continue to grow and in recent years there had been an exponential growth in the number of higher technical institutions and the number of graduates coming out of the technical institutions in India (especially in the private self supporting sector). On the outset it looks encouraging and impressive, however the real picture is revealed only when one looks at the quality of output from such institutions. Quality is an inherent or distinguishing characteristic which identifies someone or something. In this paper the issues and challenges faced by the technical education sector in India in the wake of the mushrooming of new generation institutions are discussed taking the Kerala scenario as an example. The issues related to inputs and teaching-learning processes which decide the quality of output are examined in detail in this paper. Quality assurance system is suggested as a possible strategy to achieve excellence.

Introduction

The social and economic changes happening in a country are strongly linked to the technological progress which in turn depends on the quality of technical education imparted. Developments in technology offer the people opportunities to achieve new goals, do things in new and better ways bringing about economic development and hence an improvement in the quality of life. The world today belongs to teams of professionals and specialists who can conceptualise, design, produce, operate, control and manage highly complex technical equipments. In order to be a successful team member one has to go through a carefully planned system of learning and training which enables him/her to acquire knowledge and skills essential to solve a host of complex problems. Technical education institutions thus must work within an appropriate carefully planned framework in order to discharge their responsibilities to the society efficiently. Since technical education determines the development and socioeconomic condition of a nation, there is a greater need for high quality technical education to produce technically skilled manpower in India. Till two decades back, engineering education has been confined to state run or state controlled institutions, number of seats have been limited and the admissions have been very competitive. The rapid developments in the fields of transportation, communication and computing have created an environment where many fold opportunities were created for engineering graduates. The limited numbers of state run institutions were unable to cater to the increased demand. In addition, since sufficient resources were not available with the State to start new technical education institutions to cater to the increasing demand, the society had to look for alternate sources of funding for meeting the increased demand which resulted in opening of institutions in private sector. Many entrepreneurs saw this as a business opportunity, which resulted in the creation of self supporting institutions which have since mushroomed.

Growth of Technical Education in Kerala

The first Engineering College in the State of Kerala started functioning in 1939. The number of engineering colleges grew from 1 in 1939 to more than 115 in 2011and seats grew from 39 to more than 32000 during the same period. In Kerala, at present we have different types of engineering colleges; viz. government owned, government-aided, private, government controlled self supporting, self supporting institutions run by private trusts, and self supporting institutions owned by individuals. The major mile stones in technical education in the State are given in Table 1.

Table 1 : Major Milestones in Technical Education in Kerala

1939	First Engineering College: CET (Government) (21 seats)
1958	First college in Private Sector (Government Aided): TKM Kollam
1961	First Centrally supported institution: REC Calicut
1989	First Self supporting College –run by a Govt. agency : Model Engg. College
1994	First self supporting College run by minority trust: MES, Valancherry
2000	Self supporting colleges by Private trusts

The Table 2 shows the different types of institutions in Kerala based on the type of management.

Table 2 : Types of Institutions in Kerala based on management

1. Government
2. Private - Government aided
3. Self supporting –Run by Government agencies
4. Self supporting – Run by Universities
5. Self supporting – Run by Minority community
6. Self supporting – Run by Trust/Institution
7. Self supporting – Run by Individual/Family trust

The growth in numbers has been phenomenal but the geographical distribution of the engineering colleges has not been even. Table 3 shows the institutions in different districts of Kerala in 2011. Thiruvananthapuram district leads wi n 21 instituions and Wayand district has only one.

Table 3 : Distribution of Engineering Colleges in Kerala

District	Number of Engineering colleges			
	Govt & Aided	*Govt. Controlled Self supporting*	*Private Self supporting*	*Total*
Alappuzha	-	3	6	9
Ernakulam	1	1	16	18
Idukky	1	2	2	5
Kollam	1	3	9	13
Kannur	1	1	3	5
Kasaragod	-	2	-	2
Kottayam	1	2	5	8
Kozhikkode	1	1	3	5
Malappuram	-	1	4	5
Palakkad	2	-	6	8
Pathananthitta	-	2	2	4
Thrissur	1	-	10	11
Thiruvananthapuram	2	4	15	21
Wayanad	1	-	-	1
Total	**12**	**22**	**81**	**115***

*In addition there is an NIT, one College of Agricultural Engg & Technology (KAU), one College of Diary Science & Technology (KVASU) and one College for Architecture (Private self supporting). In the year 2012, the number of institutions has risen to 130 and list is incomplete (as per the data available from CEE, Kerala).

Ever since the number of institutions increased, the issue of quality became a major concern. The ultimate quality of any system directly depends on the quality of inputs to the system and the excellence of the processes under taken to produce the output. It is with this in mind that one has to examine the quality of education in newly formed engineering colleges in Kerala.

Definition of Quality

There are different definitions for quality. The Oxford dictionary defines quality as *"The degree of excellence of something as measured against other similar things"*. In Webster's dictionary it is explained as *"Character with respect to excellence, fineness etc."* and *"Grade of excellence"*. The synonyms for quality include *'The best'*, *'Consistency'*, *'Value for money'*, *'Customer's delight'*, *'Compliance with standards'*, *'Conformance to requirements'*, *'Compliance with specifications'* etc.. The differing definitions cause confusion or conflict. In general we can use the following terms to signify different aspects of quality;

- ***Grade*** - to indicate degree of excellence
- ***Quality*** – to refer conformance to requirements
- ***Fitness of Purpose*** - to represent the degree to which a set of inherent characteristics fulfil requirements (ISO 9000-2000)

Quality in Engineering Education

The quality in education is a complex topic with varying conceptualizations. The Quality education from a TQM perspective is; *"total quality management in education is multi-faceted – it believes in the foundation of an educational institution on a system approach, implying a management system, a technical system and a social system..It includes within its ambit the quality of inputs in the form of the learning and teaching activity; and the quality of outputs in the form of enlightened students that move out of the system"* [4].

It can be defined with the following parameters;

- Excellence in education

- Value addition in education
- Fitness for purpose
- Fitness of educational outcome and experience for use
- Conformance of education output to planned goals, specifications and requirements
- Defect avoidance in education process and Meeting or exceeding customer's expectations of education.

The engineering education is a sub set of education. The US National Science Foundation (NSF) Task Force on TQM has come up with the following definition of Quality Engineering Education: *"Quality Engineering Education is the development of intellectual skills and knowledge that will equip graduates to contribute to society through productive and satisfying engineering careers as innovators, decision makers and leaders in the global economy of the twenty first century"*. Quality Engineering Education demands a process of continuous improvement of and dramatic innovation in student, employer and societal satisfaction by systematically and collectively evaluating and refining the system, practices and culture of engineering education institutions [5].

Components of a Technical Institution

The basic components of a technical institution are the students, the infrastructure, the teachers, the curriculum, the teaching and learning aids[1], the linkage mechanism with industry-institute and other user system, the management system, the support services system, the guidance and counselling, the internal and external evaluation system, the feedback system [2] etc.. There are other important components which are called the process components. They are, way of teaching, the way the students learning, students activities beyond the regular time table, the motivation of both faculty and students, attitude of the management, the overall academic climate, the opportunities and encouragement for innovations and creativity, research and development, the openness of communications, the leadership qualities of head of institutions and of departments, the sense of involvement in providing quality services, the organization structure, the quality of team work, the reward and recognition system, the faculty development programme, the appraisal system, the clarity in the vision and objectives of the organization.

Indicators of Quality

While examining quality of engineering institutions one must know what to look at. The indicators can be for different components such as students, faculty and institution as a whole. Following are the usual indicators of the quality.

Indicators of student quality:

- Number of students completing degree
- Time taken to complete the course
- Proportion undertaking practical training
- Proportion participating in research and development
- Employment profiles and salaries on Graduation
- Number of students recruited by reputed companies
- Number of students seeking post-graduate studies
- Satisfaction levels of students and Employees
- Perceived reputation of graduates and alumni, nationally and internationally
- Number becoming entrepreneurs
- Passing percentage of the students with higher class

Indicators of faculty quality

- Number of applications for faculty position, at different levels
- Academic quality, in terms of publications, honours, awards, patents, sponsored projects and consultancy
- Retention success; turn-over
- Teaching quality, innovative initiatives
- Publication records
- Sponsored research, consultancy and continuing education activities
- Professional society and public service involvement
- Ability to mobilize resources for department and institution
- Internal and external (national and international) honours and awards
- Quantum of practical experience
- Effectiveness of student counselling
- Faculty career satisfaction levels

Indicators of institutional quality:

- The utilization of strategic planning Processes
- Interaction with the environment, industry, profession, community
- Mobilization of resources for institutional development
- Diversity of external financial support
- Demand from outside agencies for R & D and continuing education
- Adjunct appointments with industry
- Inter-disciplinary activities
- Self-assessment and accreditation processes
- Alumni involvement
- Perceived reputation, nationally and internationally
- Use by national agencies as think tanks and for technology development
- Leadership in education and research

Assessment of New Generation Institutions

Recently the honourable High court of Kerala directed the Government of Kerala to constitute an expert team to carry out surprise inspections in self financing Engineering Colleges in the State to verify physical availability of faculty and co-relate the staff position with the declared strength for university affiliation. The committee constituted for the purpose have found that the infrastructure facilities of majority of new generation institutions are satisfactory. On the other hand regarding faculty, the findings were really shocking. In a large number of institutions even the minimum number required as stipulated by the AICTE (faculty student ratio of 1:15) is not present. The required minimum qualification or the cadre ratio is far from satisfactory. The pay packages offered are far below the AICTE packages. The faculty with sufficient experience and research background are a rarity. In many institutions the faculty consisted of retired professors from government/ aided institutions and a group of fresh graduates. There are no professors in many departments, no middle level faculty members and technical support staff.

The result analysis of the students in the university exams have revealed that in all universities the overall pass percentage is below

50%. One of the reasons for this poor performance is admitting students in such institutions because of parental compulsions rather than genuine interest in engineering subjects. The teaching-learning processes in majority of new generation colleges do not go beyond the school culture and standards. The innovation and creative abilities and intellect of the students are never challenged. They are trained to memorise equations and essays and reproduce it in university examinations. The research ambience and the project culture is totally absent.

Quality Assurance

Quality assurance refers to the planned and systematic activities implemented in a system so that the quality requirements for a product or service will be fulfilled. It is based on the principle that prevention is better than detection and cure. While the so called inspections will help one to detect the problems, it is not a remedy by itself. Quality control measures are focussed on process outputs and meant as a cure. Quality assurance is the systematic measurement, comparison with a standard, monitoring of processes and an associated feedback loop that confers error prevention. Quality assurance focuses on enhancing and improving the process that are used to create the end result, rather than focusing on the result itself. Conventionally the quality assurance has been practised in manufacturing industry with great success. It involves the following basic principles;

- Customer Focus
- Leadership
- Involvement of People
- Process Approach
- System Approach
- Continuous Improvement
- Factual Approach
- Mutually beneficial Supplier Relationship

Quality assurance verifies that any customer offering is produced and offered with the best possible materials, in the most comprehensive way and with the highest standards. Quality assurance also helps determine whether the steps that are used to provide the product or service are appropriate for the time and conditions. Quality assurance in the conventional sense demands a degree of detail in order to be fully implemented at every step. The goal to exceed customer expectations in a measurable and accountable process is provided by quality assurance. But in the context of technical education scenario, institutions are handicapped by factors beyond their control especially the case of inputs.

In order to implement a quality assurance strategy, there are 4 distinguishable steps, often called a PDCA cycle. During the first step of the cycle, Plan, the institution should establish its objectives and determine the processes or changes in the processes that are required to deliver the desired results. But during planning one can include determining specific levels of quality or measurable results that the institution wants to achieve. The second step, Do, is when the processes or changes are developed and tested. In the third step, Check, the processes or changes are monitored and evaluated to determine whether the results are meeting the predetermined objectives. Checking could involve testing and other objective measurements to determine whether the goals were met. The final step, Act, is when actions that are necessary to achieve the desired improvements are fully implemented into the process. The cycle can then be repeated, beginning with new objectives being planned.

The following are the major elements to be considered in the development of a QA system to be implemented.

- **A Core team responsible for coordinating QA activities**: This will not be a problem if the management has a commitment to achieve quality
- **A QA plan** : Here the institution has to set its objectives (based on pass percentages, core competencies of the students, the placements achieved by students or feedback from the customers)
- **General QC procedures**: The general QC procedures will look at the achievements at the macro level for the institution as a whole viz. Accreditation, ISO certification etc.
- **Specific QC procedures** : Here one has to look at micro level QC requirements for all elements such as students, faculty, infrastructure, support services (Students : pass criterion, screening test limits assignments, lab work, reports, mini projects, seminars, presentations, attendance, discipline, etc., Faculty : qualification improvement, subject upgradation,

training programmes attended, papers presented, consultancy and research work under taken etc.,)

- **QA review procedures** : periodic review procedures of quality standards to be used have to be set in advance.
- **Reporting, documentation, and archiving procedures**

Conclusion

Quality is a major concern in the competitive environment where the 'Survival of the fittest' is the order of the day. If the institutions do not take necessary steps to maintain quality of the output they are prone to lose out. In this paper a Quality Assurance strategy is suggested as a possible answer to the quality issues of technical education institutions. Quality assurance of students, faculty, infrastructure and support services have to be addressed. A committed management is the primary requirement for the success of implementing this strategy. A Plan-Do-Check-Act (PDCA) Cycle of QA will help institutions stay in the race and achieve excellence.

References

1. Sirvanci M. B., "TQM implementation: Critical Issues for TQM implementation in higher education", the *TQM Magazine*, Vol.16, No.6, 2004, pp. 382-386.
2. R. K. Jain, Pravin R. P., Viranchi C. P, "An analysis of student's feedback on current scenario of teaching-learning process" The Indian Journal of Technical Education, Vol. 31, No.4, 2008.
3. Lidia Z. Akowska,"Engineering education: from European to polish perspective", Engineering education-perspectives, issues and concerns edited by Rajarshi Roy, Shipra Publication, 2009, New Delhi.
4. Sangeeta Sahney, Devinder Kumar Banwet, and Sabita Karunes, "Enhancing quality in education: application of quality function deployment-an industry perspective", Work study, Volume 52, Number 6, 2003, pp. 297-309.
5. R. Natarajan, "*The Role of Accreditation in Promoting Quality Assurance of Technical Education*", International Journal of Engineering Education, 2000, Vol. 16, No. 2, pp. 85-96.
6. Ram Prasad, "Indian education sector challenges ahead", The Indian Journal of Technical Education, volume25, No.4, 2002,

pp-73.

7. Sudarsanam, "Human values in technical education", The Indian Journal of Technical Education, volume25, No.4, 2002, pp-1.
8. Veeranna D. K. et al., "Quality issues in Engineering Education : An Indian Perspective", International Journal of Engineering Science and Technology (IJEST), Vol. 3 No. 3 March 2011, pp-2044.

❑❑❑

3

Role of Modular Courses/ Finishing Schools in filling up the Gap Between Education and Employability

**Shakila Shamsu*

ABSTRACT

Tertiary education is a critical pillar of human resource development which provides not only the high-level skills necessary for every labour market but also the training essential for all professions. Currently, India is passing through an unprecedented phase of demographic change. The ongoing demographic changes are likely to contribute to an ever increasing size of labour force in the country. The Census projection report shows that the proportion of population in the working age group (15-59 years) is likely to increase from approximately 58% in 2001 to more than 64% by 2021. But the overall population is not the issue - the proportion of population in the working age group of 15-59 years will increase from 57.7% to 64.3%. To put it in another

* Shakila Shamsu, OSD,,Department of Higher Education, MHRD, Govt. of India

way, the number of those in the 15-59 age-groups would have increased by about 308 million during the period. The large numbers of the 15-59 year olds would also be reflected in the workforce. It is estimated that by about 2025 India will have 25% of the world's total workforce. But beyond 2025 the numbers of the aged will begin to increase even more dramatically, and consequently the window of opportunity is between now and 2025.

To tap the demographic dividend, India needs better education and skills indicators and a much better labour ecosystem. A huge challenge is to train a large number of youth in the skills which are required by the industry. The government has been attempting to improve basic education through a host of measures in recent decades and achieving some success. But, vocational education and imparting skills remains a critical area of concern.

This paper attempts to outline the strategies for enhancing the employability factor of our students in the coming years through a varied range of modular and skill based courses. The inter-linkages between skills and employability and the need to have add-on courses with the existing curricula of technical and general education programmes to bridge the widening gap between education and employment will be covered.

One major concern is the lack of employable skills in our technical education students. In the XII Plan, this area is being accorded a high priority as it will critically impact the development strategies. We are looking at the country's large youth population as an advantage point. But in order to leverage this demographic factor, we need to ensure that our youth are empowered with the right skills to meet the challenges of knowledge based market economy. Based on the identification of the skill gaps in different sectors, possible approaches, such as, setting up of finishing schools, offering courses for enhancing employability are some of the strategies proposed. The XII Five Year Plan should aim at creating productive, gainful and sustainable employment opportunities with decent working conditions; therefore, 'Skills led employment' generation is a most sustainable employment generation strategy to address the needs of the technology driven employment market in the coming years.

While in the long run, skill based courses and vocational courses must be brought into the fold of mainstream education, suitable interim measures/ schemes must be initiated to address the need for enhancing the employable skills of our current technical students. Some strategies for the Employability factor, which are either faculty-centric or student-centric are: Industry Institute Student Training Support, Industrial Challenge Open Forum, Long Term Student Industry Placement Scheme, Industry-Institute Continuous Interaction Scheme, Intensive Interaction –Train The Teachers, Industry Training Programme And Support Scheme, Centre for Qualified Manpower, National Employability Portal, Tax Benefit for Teaching Laboratory Support. Some strategies for the Skills component with focus either on institution or students are: Skills Requirement Assessment Revitalizing the Diploma Education, Lateral Entry of ITIs to Diploma Programme, Vocation based Certification Programme, Industrial Finishing Schools, Bridge Courses for Skill Enhancement, Special Manpower Development Programme.

A mapping of Class X & XII has been done by the AICTE and a comprehensive scheme is being drawn up for strengthening skills and developing vocational education by utilizing the idle infrastructure of existing institutions and building in an industry-institute interface. There is a trend in some states to set up vocational universities. This is not a desirable trend as it encourages a unitary dimension of learning which goes against the concept of inter-disciplinarity and the holistic nature of knowledge. Existing technical institutions must be encouraged to develop skill based courses and vocational courses which are modular in nature with multiple entry and exit options. Since the National Vocational Education Qualifications Framework (NVEQF) is the proposed overarching academic architecture for vocational education, all initiatives in skill competency and development must fall within its ambit and suitably designed for effective implementation.

Introduction

Tertiary education is a critical pillar of human development which provides not only the high-level skills necessary for every labour market but also the training essential for all professions. Currently, India is passing through an unprecedented phase of demographic changes. The ongoing demographic changes are likely to contribute to an ever increasing size of labour force in the country. The Census projection report shows that the proportion of population in the working age group (15-59 years) is likely to increase from approximately 58% in 2001 to more than 64% by 2021. But the overall population is not the issue - the proportion of population in the working age group of 15-59 years will increase from 57.7% to 64.3%. To put it another way, those in the 15-59 age-group would have increased by about 308 million during the period. The large number of the 15-59 year olds would also be reflected in the workforce. It is estimated that by about 2025 India will have 25% of the world's total workforce. But beyond 2025 the numbers of the aged will begin to increase even more dramatically, and consequently the window of opportunity is between now and 2025.

To tap the demographic dividend, India needs better education and skills indicators and a much better labour eco-system. A huge challenge is to train a large number of youth in the skills which are required by the industry. The government has been attempting to improve basic education through a host of measures in recent decades, achieving some success. But, vocational education and imparting skills remains a critical area of concern.

This paper attempts to outline the strategies for enhancing the employability factor of our students in the coming years through a varied range of modular and skill based courses. The inter-linkages between skills and employability and the need to have add-on courses with the existing curricula of technical and general education programmes to bridge the wide gap between education and employability skills will be covered.

Approach Paper to XII Five Year Plan

It would be relevant to examine this issue as envisaged in the XII Five Year planning process. In para 10.24 sub- section **"Enhancing Employability"** within the Chapter on **Education and Skill development** of the **Approach Paper to the XII Plan** states as quoted " There is a need for a clear focus on improving the employability of graduates. Indian higher education is organized into 'General' and 'Professional' streams. General education which is an excellent foundation for successful knowledge based careers, often fails to equip graduates with necessary work skills due to its poor quality. On the other hand, professional education is often expensive, lengthy and usually imparted in narrowly specialized private institutions, with little emphasis on liberal arts, which is essential for the development of intelligent able-minded citizens. For both 'General' and 'Professional' education streams, integrated curriculum with greater flexibility in choice of subjects and innovative pedagogic practices are needed to improve the quality and hence employability. Graduates now require the skills beyond the basics of reading, writing and arithmetic (the '3Rs'). Skills such as critical thinking, communication, collaboration and creativity (the '4Cs') are now important in more and more jobs. Accordingly, there is need to focus on the '4Cs'. Special emphasis on verbal and written communication skills, especially in English would go a long way in improving the employability of the large and growing mass of disempowered youth."

Royal Academy of Engineering Report, UK

The Royal Academy of Engineering Report UK has identified, after detailed research, six specific skills which are of particular significance to employers, who are employing engineering graduates. These have been listed as:

- Practical application of theory to real industrial programmes
- Theoretical understanding
- Creativity and innovation
- Team working
- Technical Breadth and
- Business skills

Based on these findings, the technical institutions in UK have created a framework which would lead to producing a highly employable graduate engineer. This framework focuses on three aspects of training which are aimed at enhancing (i) technical competence (ii) commercial awareness (iii) soft-skills development. Each of these aspects needs to be analyzed separately.

Under technical competence, the students grasp of cognitive knowledge, engineering science, mathematics and his/her core subjects become important. Similarly his/her core competency in design analysis, materials, manufacturing processes and skills required for enhancing competency in these technical areas is critical. The cognitive knowledge combined with technical skills provides the substance of technical competence that is expected in an employable engineering graduate. This technical competence needs to be complemented with high degree of commercial awareness. Commercial awareness needs to be provided to engineering students through Business Management studies, visits to companies and business houses having guest speakers from industry, student placement into companies and programme sponsorship into commercial establishments. Having technical competence without awareness of financial and business environments has proved that engineering graduates fail to become good managers. The third important component essential to mould an engineer into an employable product is ensuring the soft skills development. What does this consist of? It includes the team working, negotiating skills, oral and written communication skills and information technology skills. This framework with the above mentioned three components have proved a way to mould a fresh engineering graduate into a potentially attractive employable candidate.

Report of the Council for Industry and Higher Education (CIHE), UK

A lot of thought has gone into identifying, the generic skills needed in engineering graduates. The Council for Industry and Higher Education (CIHE) of UK in its report on graduate employability reveals five most important skills and capabilities which are sought by nearly 80% of employers/graduate recruiters. These are communication skills, team working skills, integrity, intellectual ability and self-confidence. Other less significant skills indicated are reliability, self discipline and motivation, flexibility, leadership and responsibility, creativity and sense of empathy. Several other documentation reports have listed out skills needed by potential employers which can be categorized as attitude (sincerity, sense of ownership and motivation), business ethics and honesty, grooming and confidence, general awareness, basic managerial skills (leadership, team work, time management) basic sales and customer service needed at entry level, domain knowledge and work experience.

Existing Approaches for enhancing employability and suggested strategies

Most of the educational institutions around the world have adopted a combination of varied approaches to ensure that their students are equipped with the different skills for entering the employment market. These include:

- Finishing schools
- Bridge courses

- Certificate courses
- In-service courses
- Beginners courses for specific set of industries or service sectors
- Skill based diplomas
- Modular courses
- Recognition of non-formal education
- Creating linear structures from ITI to diploma to undergraduate levels as envisaged in proposed vocational education framework.

Several suggestions have emerged, in the context of formulating the XII Five Year Plan, on changes in engineering education. It has been mooted that instead of branch specific engineering education which becomes too narrow, there could be division specific engineering degrees. Bringing in industry experts for teaching in engineering institutions and creating a reciprocal arrangement between academic institutions and industries by academics undertaking R&D and innovative projects in industry, needs to be encouraged.

Graduate employability needs to be built into the curriculum development process. Employability strategies have to be made an integral part of teaching learning to ensure that students have the opportunity to equip themselves with a full range of employability skills. In the UK, Higher Education Funding Council for England (HEFCE) has been encouraging institutions to clarify their strategies for employability, especially for prospective students, by publishing employability statements.

It would be equally pertinent to examine the broad man-power projections made by each sector of the economy. As part of the planning process for the XII Plan, every sector of the economy has been assessing the requirements of human resources, both in terms of specialization and the likely numbers required for the coming five to ten year period. Academic regulators, such as, the UGC and AICTE have also undertaken man-power forecasting. Further, each sector has been identifying specific traits, skills and expertise required for different entry levels of jobs. The professional associations and bodies also need to generate data about the estimated numbers of professionals required in the hierarchical or laddered structure of professional positions. It is strongly advisable that some sort of tripartite arrangements should be made between academic institutions, industry and the potential student aspirants.

The manufacturing sector has indicated a skill requirement of approximately 90 million across different sub sectors in 2020. The sector skill report of the manufacturing sector has provided details of the expected skills gap for the period covering 2008 to 2020. In the manufacturing sector, the data related to expected skill gaps are available in the public domain under the National Skill Development Corporation (NSDC) website. The relevant table is reproduced. The Working Group on Industry for the XII Plan has also identified four big challenges plaguing skilling in the manufacturing sector as shown below.

India's Manufacturing Sector has a skill requirement of ~90M across different sub-sectors in 2022

Sectors	Expected Skills Gap in millions			
	2008	2012	2018	2022
Textiles	28.3	34.1	42.5	49.3
Food Processing	4.7	5.72	7.9	9.8
Auto & Auto Components	2.9	4.5	6.1	7.6
Leather & Leather Goods	2.5	3.6	5.3	7.1
Gems & Jewellery	2.2	3.1	4.8	6.1
Cons. Material & Building Hardware	0.7	0.9	1.6	1.6
Electronics & IT Hardware	0.3	0.5	1.0	1.5
Pharmaceuticals	0.4	0.6	0.9	1.2
Furniture	0.3	0.5	0.8	1.1
Chemicals & Petrochemicals	0.6	0.6	0.7	0.8
Total	42.9	54.2	71.6	[illegible]

Assumptions
For the Auto Sector, proportionate growth rate assumed between 2012 and 2018
- Incremental increase in skilled manpower required in every period = CURRENT PERIOD – PREVIOUS PERIOD
Source: Sector Skills Reports www.nsdcindia.org

Skilling India's Manufacturing Sector
The Four Big Challenges

Availability of a Skilled Workforce

- Less than 25% of the total engineering graduates are employable

- Only 5% of the total Industrial workforce Is skilled

A Cogent Public Information System

- India needs a coherent LMIS to channelize its manpower resources efficiently

- Data on public domain needs immediate up-dation for devising effective future policies

Linking Jobs with Skills

- A mechanism to ensure better employment opportunities for the trained workforce

- Active cooperation between the industry and government for skill mapping and development of market driven courses

Infrastructure Development

- Equipping employment exchanges to play a more effective role for linking the industry with thelabour force

- Equipping training institutes like the ITIs in expanding training capacity and improving the quality of teaching

The other industry sectors such as, ship building and ship repair, fertilizers, steel, cement, mineral exploration and development, leather, paper, food processing, defence equipments, aero space etc. are all needed to be explored for understanding the estimated skill gaps for the respective sectors.

Suggested strategies for enhancing student employability

The Working Group on Technical Education for XII Plan under the MHRD has identified skills and employability as one of the critical areas in technical education. Consequently, a Sub group on skills and employability was constituted within this Working Group. Some of the suggested strategies are outlined below: These strategies can be broadly classified into student centric strategies and faculty centric strategies. I would like to point out that these strategies and arrangements are only suggestions and not operational or fructified. However, since they have emerged through a collaborative thought process of academia, industry and research institutions, it would be worthwhile to examine them. These could be a resource pool of ideas which could be considered for implementation by institutions or State Governments.

The Student-centric strategies are :

1. Industry Institute student training support
2. Industrial challenge open forum
3. Long term student industry placement scheme
4. Industry training programme and support scheme
5. Centre for Qualified Manpower
6. Industry Institute continuous interaction scheme

The Faculty-centric strategies are:

7. Industry Institute continuous interaction scheme
8. Intensive interaction-train the teachers

The broad contours of these strategies are outlined below.

Industry Institute student training support

The broad objectives of this strategy are:

(i) To connect industry directly with students through training programmes

(ii) Such initiatives to operate in their specific regions or specializations

(iii) Training to be imparted in every District Headquarters on regular and need based pattern

(iv) Should not create any new physical or institutional setup

The partner agencies of the Industry Institute Student Training Support could be:

- Local Industries, Each District Industrial Centres, Industrial Associations
- Public or potential employees
- Can be done jointly with concerned State Government Ministry of Industries

Industrial challenge open forum

The broad objectives of this Industry Challenge Open Forum strategy are:

(i) To directly promote industry institute interaction for industrial problem solving

(ii) Around 200 open industry problems to be floated every year through their nodal or representative agencies

(iii) Each problem is officially funded to 10 different groups intending to work on these problems throughout the country

The partner agencies of Industrial Challenge Open Forum are :

- All industries and their representative forums or associations (FICCI, NASSCOM, ASSOCHAM, CII)
- All educational institutions in the chosen areas of specializations
- State Government Ministry/Dept. of Science and Technology

Long term student industry placement scheme

The broad objectives of this strategy/scheme are:

(i) Long term project placement would bring better synergy between industry and institutes allowing them a better understanding and possible outcomes

(ii) 5-10% of the students to be placed in interested industries for at least one semester. Initially 400 select institutions with 25 students each

(iii) Both industry and institute will be incentivized through this scheme.

(iv) Mentoring from both sides to be made mandatory

The partner agencies of Long term student industry placement scheme could be approved institutions and industries (PSUs, Large/ Heavy Industries, SSIs etc.), GoI supported Labs (CSIR, DRDO, ICAR etc.) in vicinity of each other (500 km radius only).

Industry training programme and support scheme

The broad objectives of this strategy/ scheme are:

(i) To enhance the industry's direct initiatives for increasing the employability

(ii) To recognize and support industries efforts

(iii) To promote and scale up such efforts and initiatives by several other industries

The partner agencies of Industry training programme and support scheme could be WIPRO-Mission 10X, INFOSYS-IUCEE, TCS, CISCO, IBM, EMC or any other organisations.

Centre for Qualified Manpower

The broad objectives of the Centre for Qualified Manpower are :

(i) To be ready with potentially qualified and identified staff for various sectors of industries or employment opportunities

(ii) NASSCOM conducting NAC-TECH for IT employment; similar efforts can be undertaken by the various other sectors

The partner agencies of Centre for Qualified Manpower are:

- Industry bodies or associations
- All individual job seekers

Industry Institute continuous interaction scheme (Student- Centric)

The broad objectives of this strategy/ scheme are:

(i) To support 4-5 industrial interaction per month (one per week)

(ii) Industrial expert spends two full days of activity in an institutional environment alongside faculty and students.

The partner agencies suggested are:

- Registered Industries, PSUs, MNCs, Industry association
- Recognized educational institutions

Industry Institute continuous interaction scheme (Faculty centric)

The broad components of this strategy/ scheme are identical to the ones mentioned under the previous strategy, except that the focus in this would be for the faculty. Like students, faculty too needs to be supported for industrial interaction at regular intervals.

The partner agencies would be registered industries, PSUs, MNCs, Industry Association and recognized educational institutions

Intensive interaction-train the teachers (Faculty centric)

The broad objectives are:

(i) To make faculty learn about the needs and environment of industry by spending a month in industry during summer vacation time

(ii) 5% of the staff encouraged and incentivized to take it up every year

(iii) Incentives could be honorarium and living expenditure

(iv) Periodic refresher courses or workshops to familiarize with changing needs

The partner agencies of the train the teachers could be:

(i) Well established and large industries, PSUs and MNC to be the host industries

(ii) All accredited institutions to be partners in the scheme limited to 1000 institutions.

Conclusion

There is a growing recognition of the need to align higher and technical education both with the aspirations of the learners and also with the economy and labour market. This has inevitably brought up the critical issue of enhancing employability skills of our fresh graduates. Various ways to move forward in this direction are being attempted by the governments, institutions and regulators. While no single approach might yield the desired results, engaging multiple strategies to suit different institutions will need to be worked out. There is a wealth of reports available in public domain which provides international and national perspectives on this eminently pertinent issue that is currently grappling the education sector across the world.

(The views expressed by the author are purely personal)

References

1. www.raeng.org.uk
2. www.cihe.co.uk
3. Working Group Report for Technical Education for the XII Plan, MHRD, www.planningcommission.gov.in
4. Steering Committee Report on Industry-www.planning commission.gov.in

4

Industry Academy Linkages in the Global Era

Issues of Employability of Engineers

*Shakila Shamsu

ABSTRACT

Globalization, a key reality in the 21st century, has already profoundly influenced higher education. Globalization can be defined as the reality shaped by an increasingly integrated world economy, new information and communications technology, the emergence of an international knowledge network, the role of the English language, and other forces beyond the control of academic institutions. Quality assurance in higher education has raised to the top of the policy agenda in many nations. Post-secondary education has to prepare graduates with new skills, a broad knowledge base and a range of competencies to enter a more complex and interdependent world.

Currently, around 500,000 engineering graduates come out of our engineering colleges. This number would cross one million in three to four years from now. There is however a serious question

* Shakila Shamsu, OSD, Department of Higher Education, MHRD, Govt. of India

about their quality. Industry finds a large number of these graduates unemployable needing further training.

A strong industry–academia relationship is of key importance not only for producing technical graduates suited for industry needs but also for creating an ecosystem of innovation and entrepreneurship. The Kakodkar Committee has made specific recommendations for encouraging industry to set up their research labs at IITs. Indian industry has been changing rapidly in the last 10 to 15 years. Today they want to become the world's best. Unlike in yesteryears when they depended entirely on import of technology, today Indian industry recognizes the necessity to carry out and participate in R&D to master their technology. Unfortunately, they have little understanding or experience in conducting research, though they may have acquired the ability to indigenize and productize technologies. Our technical education institutions need to develop strong interfaces and linkages with industry to meet their technological needs. This would be advantageous from the perspective of making our engineering graduates acquire the relevant skills and knowledge required by the industry.

For the XII Plan, there is a proposal to promote innovations by expanding the Technology Business Incubation Infrastructure. Some of the better institutions like IITs have created technology business incubation facilities in their campuses, which are proving to be focal points for promoting innovations amongst students and faculty, to consider working towards taking some of their applied research to the market through creation of business models for the same. These efforts need to be expanded greatly (a) by scaling the previously successful centres of such innovations, and (b) by creating many such centres across the higher technical institution in the country.

This paper attempts to examine the contemporary global demands of technical education and the critical need to make our engineering graduates aligned to them.

Overview

The industry academia linkage is being advocated so that the relevant stakeholders, both institutions and individuals, gain from the partnership. Academic institutions, such as, universities, colleges and technical institutions as well as research based institutions would immensely benefit from the projects and other academic tie-ups with industry and corporate houses. The individual stakeholders of this academic segment, namely, the students, faculty and researchers would have an enriched experiential learning advantage through their deeper interaction with industries and corporate houses. As far as the industry is concerned, their research needs creation of patents and other innovations which can be propelled through the academic institutions and their faculty, students and researchers. Such a mutually beneficial arrangement has become more significant in the recent times particularly due to the constantly changing socio-economic needs of an industrialized and knowledge based society.

While promoting industry academia linkages to enhance the quality of learning experience as also to improve the learning outcomes of students, the goals could be varied. One end of the spectrum would be to enhance the employable skills of graduates, whereas, at the other end, would be to promote research & development, innovation and entrepreneurship. At the basic level, the industry could gain from equipping the undergraduate students with the right attitudes, skills and application based knowledge to suit their requirements.

Industry academia interface can be described or defined as an interactive and collaborative arrangement between academic institutions and business corporations for the achievement of certain mutually inclusive goals and objectives. It has been generally felt that the successful products of our educational system, be they general education graduates or technical education graduates, will be better suited to industry needs if there are institutional mechanisms for promoting industry academy interaction within their curricula.

Why the need?

It would be relevant to understand the various reasons put forward as the rationale for greater industry academy interaction, particularly in the case of engineering students. Some of major reasons often identified are :

- Constantly changing needs of the industry
- Increasing criticality of human competence in creating and sustaining competitiveness
- Increased competition for students placements

- Growing need of industry to make their fresh recruits productive with all the right knowledge, skills and attitude thereby reducing the training costs.
- Increasing interdependence between academia and industry to satisfy need for sustenance and innovation in their respective areas.

Generally, there is a tendency to advocate industry academic partnership only for students in technical institutions. Though the current seminar is addressing the engineering segment, it is in my view a very restrictive perspective as there is a critical mass of students who are pursuing their non-technical courses in arts, commerce, basic sciences, social sciences, humanities etc. and it is equally, if not more, important to ensure that these students are also properly attuned to the business, industrial and economic employment markets. Presently, India has a pre-dominant advantage over other countries, in so far as it has a substantial youth population. This so called demographic dividend is not confined only to students pursuing technical education, such as those in engineering, management, pharmacy, architecture, hotel management etc but other academic disciplines of tertiary education. Hence, to address the issue of industry academia linkages from the point of technical education alone would be highly myopic. To cite an example, students pursuing Bachelor of Arts have scope to take up jobs in the services and tourism sector. There is lot of gain if such students while pursuing their under-graduation are rightfully placed for summer projects or some other academic arrangement in relevant industry so that the relevant applications and skills, such as, communication skills, personality development and soft skills are provided as value-added knowledge. This would facilitate their being job ready as soon as their education is completed and employers will also be equally happy to pick up such candidates who would not need further training and thereby save their time and training costs.

Strategies and Methods

Various strategies and methods can be employed by educational institutions offering technical and tertiary education programmes for greater collaboration with industry. Some of them which are in vogue are:

- Guest Lectures
- Joint seminars by academia and industry
- Consultancy
- Academia generating ideas and acting as incubators to new business.
- Training and Internship of students
- Industry representation in Governing Councils and Board of studies
- Executive Education/Management Development programmes
- Industry inputs in curriculum designing
- Industry providing financial and infrastructure support
- Funding academic and applied research
- New emerging model of student mentoring
- Setting up knowledge clusters
- Industry to earmark a part of the surplus for activities of preparing students

Benefits of the Linkages

The interaction between the students of academic institutions and industry houses is mutually beneficial and has valuable spin offs for both of them. In the long run, the institution gains by the rich experience of various sector based industries and their productive demands. Similarly, the industry can treat the educational institutions as incubation centres for their research and development thereby reducing the cost of R&D. On the human resources side, the quality of both teachers and students substantially improve with the infusion of industry experience and research and development. The inter-mix of the two results in greater innovation, increase in entrepreneurship and more patents and greater intellectual property rights.

The benefits for the academic institution are:

- Real time R&D in the institution
- Hands on R&D experience for students and faculty
- Training of students and faculty
- Better Teaching
- Possible IPR rights

- Possible Patents

The benefits to the industry are:

- Extension Centre for R&D efforts
- Greater access to trained researchers and teachers
- Greater access to student workforce who's potential to excel can bring in ideas.
- Lower cost of R&D
- Greater IPR Rights

AICTE XII Plan Proposal

The All India Council for Technical education (AICTE) has proposed a model for promotion of research through industry institute interaction for the coming Plan period. It proposes to establish one research park in one of the state engineering colleges in each state so as to promote industrial research in a collaborative manner between the academia and industry. Under this model, it is proposed that:

i) About 100 institutions to be selected within the country based on certain stipulated criteria

ii) The selected institutions to provide physical space on the campus for setting up of the R&D Centre of the Industry.

iii) Identify Industries in and around the location of the institution for setting up of R&D Centres.

iv) Industries to be persuaded to set up R&D Centres in the selected institutions.

v) Funding to be provided to these institutions based on defined outcomes.

Other Proposed initiatives for promoting research and innovation through industry academy linkages.

- Create a Special Empowerment Fund for supporting academia-industry-Research lab Interaction
- Provide incentives to industry to support research in academic institutions.
- Virtual Research Academy: To help promote joint research across institutions and between institutions and industry.
- Substantially Expand Existing Technology Business

Incubation Units (TBIU): Support such Centres across the country with at least 50% of the expenditure reserved for "venture funding".

- Set up at least 100 new TBIU's or Innovation Centres on a competitive basis during the XII Plan.

While the proposed initiatives can be examined by the educational institutions for implementation, it would be informative to share an existing industry academy initiative that is ongoing in the manufacturing sector. The initiative is referred to as the Visionary Leaders for Manufacturing (VLFM) programme. The project has originated as a bilateral arrangement between the Governments of Japan and India with support from Japan International Cooperation Agency (JICA) and National Manufacturing Competitiveness Council (NMCC). It is considered as the first worldwide innovative manufacturing management transformation programme created with the guidance from Japanese experts. The objective of this programme is to create a community of 500 visionary leaders to lead the Indian manufacturing centre. It is a unique partnership between industry, academia and Government. The partnering agencies are the GOI, the Confederation of Indian industry (CII), JICA IIT's, Kanpur and Chennai and IIM, Kolkata.

Innovation

Innovation is defined as the creation of better or more effective products, processes, services, technologies, or ideas that are accepted by markets, governments, and society. Innovation differs from invention in that innovation refers to the use of a new idea or method, whereas invention refers more directly to the creation of the idea or method itself. To innovate is to make changes in something established, especially by introducing new methods, ideas, or products.

Education is undeniably the critical enabler for developing the required intellectual capital which could unleash India's innovation potential. Innovation, Education and Entrepreneurship are inter-related circles within a triangle, each of which mutually influences and impacts the other. Innovation must be ingrained into all sub-sectors of education ranging from vocational education and skill development to research level.

Internationally, there are several models for setting up innovation centres in engineering universities and institutions. Every

University/Engineering College should encourage the students to initiate student innovation centres where the students and the faculty will be encouraged to develop innovative models, products, processes and even business models. Every University/Engineering College Innovation Centre will have a link with the innovation centres located at the Engineering/Science Institutions in the neighbouring areas.

The University should be encouraged to start Intellectual Property Cell which will look into the possibilities of patenting the innovations by the students/teachers of the university. In case of innovations, which will have a global impact in the Technological Domain, there should be a provision of filling patents even at the international levels, such as in U.S.A.

Entrepreneurship

A higher end objective of industry academia interaction is instilling and encouraging entrepreneurial skills in our young graduates. Entrepreneurship can be defined as the professional application of knowledge, skills and competencies and/or of monetizing a new idea by an individual or a set of people by launching an enterprise *de novo* or diversify from an existing one, to pursue growth while generating with employment and social good. This is distinct from seeking self-employment as in a profession or in a trait.

Recent surveys, such as those undertaken by Goldman Sachs and Price Water house Coopers; have estimated that India has the potential to be among the world's leading economies by 2050.

The National Knowledge Commission in its Report on entrepreneurship has referred to the Entrepreneurship 'Pyramid' in India (in terms of sectors and numbers of people engaged):

- Level 1: Agriculture and other activities: Crop production, Plantation, Forestry, Livestock, Fishing, Mining and Quarrying.
- Level 2: Trading services: Wholesale and retail trade; Hotels and restaurants.
- Level 3: Old economy or traditional sectors: Manufacturing, Electricity, Gas and Water supply.
- Level 4: Emerging sectors (including knowledge intensive sectors): IT, Finance, Insurance and Business services,

Construction, Community, Social & Personal Services, Supply Chain, and Transport-Storage-Communications etc.

At a general level, the multi-dimensional nature of the required entrepreneurial skills originates in education and involves a combination of critical (objective, analytical and logical) as well as creative and empathetic (lateral, imaginative and emotional)thinking. Imparting such skills is a process, which starts right from the school stage. However, critical aspects of higher education include professional education, vocational Education and Training (VET) and skill development.

Suggested Strategies to Enhance Entrepreneurship

1. Content and Learning Methods: NKC in its consultations with entrepreneurs found that it is important to develop pedagogic methods, examination systems and curricula that encourage critical and lateral thinking. The ability to promote entrepreneurship requires an enquiring mind that is able to make connections between theory and practice. Entrepreneurs feel that there is a need to recreate real-life situations in the classroom with the help of examples and illustrations so that students are able to get a 'concrete feel' of various aspects of the outside world. Pedagogy should encourage original thinking and not rote learning and this needs to start right from schooling stage.

2. Teaching and Entrepreneurship: NKC has also suggested that while entrepreneurship has been introduced as a course in a number of business school curricula, there needs to be a greater thrust in making entrepreneurship an important core subject. A recent initiative is to set up a School of Entrepreneurship offering undergraduate and post-graduate degrees in IIT Kharagpur. Such initiatives need to be encouraged for widespread adaptation in other educational institutions, including institutions located outside metropolitan areas. The Indian Diaspora and alumni networks need greater encouragement in funding such departments.

3. Encourage Student-led Entrepreneurial Activities on Campus: Entrepreneurship cannot be learnt by 'chalk and talk' and shared activities outside the confines of the classroom are imperative to develop leadership and team building skills that are necessary for entrepreneurship. Therefore, it becomes necessary to encourage activities within educational institutions, where entrepreneurial abilities find opportunities and fertile ground to grow. In several educational institutions, student led and faculty supported activities have been able to create awareness and prepare necessary mindsets about entrepreneurship. Examples include the Entrepreneurship and Innovation Cell at IIM Bangalore and the Entrepreneurship Cell at IIT Bombay, which engage in activities such as business plan contests, assisting start-ups and incubation centres. Initiatives such as 'business plan contests' (with angel/ VC/private sector assistance, which could directly become entrepreneurial ventures with rewards in terms of seed funding), instructional programmes that deal with the basic practicalities of entrepreneurship and platforms to interact with other entrepreneurs and the financial community (especially VCs and angels) need to be intensified and encouraged.

Way Forward

During the XII Plan, it is proposed to have greater focusing on Research and Innovation. There is a proposal to promote innovation by expanding the Technology Business Incubation Infrastructure. Some of the institutions like IITs have created technology business incubation facilities in their campuses, which are proving to be focal points for promoting innovations amongst students and faculty, to consider working towards taking some of their applied research to the market through creation of business models for the same. These efforts need to be expanded greatly (a) by scaling the previously successful centres of such innovations, and (b) by creating many such centres across the higher technical institution in the country.

Developing an eco-system of research and innovation through Design Innovation Centres and Innovation Platforms in our existing Universities is another thrust area proposed in the XII Plan for which consultations are ongoing with the National Innovation Council. While some of IITs already have research parks and design centres, it is proposed to institutionalize innovation centres in Central universities and institutions. Research promotion in the field of social science is to be given a major boost.

Educational institutions have to move beyond the phenomenon of 'working with employers' towards the concept of 'working with partners'. The collaboration between the two is dynamic and complex. A synergistic relationship has to be carved between the institutes and the industry so that both can benefit and also contribute to enhancing the quality of pedagogy.

Conclusion

This paper has attempted to examine the growing need of collaboration between academic institution and industry in the current economic scenario. Educational institutions must collaborate and interact with industry so as to empower their students to meet the requirements of various industrial sectors. Industry also needs to take a more pro-active role in setting up incubation centres, research parks and innovation centres in academic institutions for promoting R&D, innovation and entrepreneurship. A transformation in the way of academic transaction takes place with greater emphasis not only of the cognitive and knowledge but also on the applications of this knowledge in real life industrial, commercial, scientific and economic enterprises will decide how employable are our successful students in their future careers. Creating the right ecosystem rests on all concerned stakeholders so that enabling frameworks and environments are suitably worked out for enriching interfaces among academic and research institutions and industry and corporate houses. This would determine the future of our youth which in turn would make our country to become a globally recognized knowledge society. We all need to work towards this goal if we are to leverage the demographically advantageous position that India is so fortuitously blessed with today.

References

1. A Report on Innovation in India, 2007, National Knowledge Commission, www.knowledgecommission.gov.in
2. Entrepreneurship in India, 2008, National Knowledge Commission, www.knowledgecommission.gov.in
3. Working Group Report for Technical Education for the XII Plan, MHRD, *planningcommission.gov.in*

5

Industry-academia Linkages and Employability of Fresh Engineers

Review of Recent Trends in India

** Dr.S.K.Chakravorty and ** Dr.K.P.Sunny*

ABSTRACT

It is a well-known fact that in India there is very little interaction between academic institutions and the industry as far as the course curricula are concerned. This lack of interaction many a times insulates the academic institutions from the fast changes that are taking place at the industrial scene and the academic courses focus mainly on theories with little practical applications at the manufacturing scene. As a result of this dichotomy, recent years witness a peculiar situation where large number of fresh engineering pass outs remains unemployed even after acquiring degrees and diplomas in various specialized technical fields. In this paper an attempt has been made to unravel this issue in detail and to find out the missing links and suggest measures on

* Dr. S.K.Chakravorty, Deputy Director General

** Dr.K.P.Sunny, Group Head (ES &SS),National Productivity Council, New Delhi

how to make the wealth of this knowledge pool useful for nation building.

It has been reported that Indian industry is experiencing shortage of qualified skilled work force in the recent years. India has witnessed substantial increase in the number of technical institutions both at public and private domains in the recent years resulting in substantial increase in the supply of technical manpower. Many of these institutions are lacking in required infrastructure and qualified teaching faculty to impart quality education. Moreover, these institutions are focusing mainly on income generation rather than imparting quality education. There is dwindling demand for technical manpower across the globe due to economic recession. However, in the domestic scene the industry is facing shortage of skilled technical manpower while fresh engineering graduates are struggling to find employment opportunities and livelihood options after spending so much for the education. Many well known academic institutions in the west closely interact with the industry especially in the research areas resulting in path breaking inventions and innovations in the production process. As a result of the lack of inter-linkages between industry and academia, India lags behind other major countries in terms of R&D expenditure, R&D personnel, number of patents awarded, scientific articles, knowledge transfer, innovative capacity etc.

Since India was economically and educationally backward in the immediate decades succeeding independence, the focus was mainly on imparting theoretical education rather than application side. Only basic engineering subjects were taught in the Universities and Technical institutions resulting in churning out students having only generic theoretical knowledge. However, the situation got changed especially in the years succeeding the economic liberalization after 1991; there is unprecedented demand for technical manpower for meeting demands of the fast growing economy resulting in mushrooming of private sector engineering colleges and the resultant employability issues in the technical job market. This paper explores the issues related to the industry-academia linkages and the need for setting up finishing schools or modular courses with the active participation of industry so that the skill shortage issues can effectively be addressed and can find a meaningful solution to employability of fresh engineering graduates and also livelihood issues.

SECTION I

Introduction

India was an underdeveloped agrarian economy with very low literacy levels and economic achievements at the time of independence in 1947. Initially, the focus was mainly to impart basic education to the teeming millions mainly based on theories rather than on applications. Technical institutions were focusing on basic engineering subjects resulting in churning out students with only generic theoretical knowledge. Over the years the Indian economy had undergone considerable transformation and it emerged as one of the fast growing global economic super powers along with China. However, the educational sector was quite slow to adapt to the fast changes that are happening in the economy and still continued with theory based pedagogy than application oriented one. The situation dramatically got changed especially after India embarked on rapid pace of economic liberalization after 1991.

Unprecedented demand for technical manpower originated from the fast growing economy resulting in mushrooming of private sector engineering colleges at various states since the government run technical institutions were so few to supply the required number of technical manpower. However, many of these newly formed technical institutions were not adequately equipped to provide the required quality of education and the pass outs started facing the employability issues in the technical job market. On the one hand there is a shortage of skilled technical manpower in the industry while on the other hand the fresh engineering pass outs are found unemployable by the industry. As a result of this dichotomy, recent years witness a peculiar situation where large number of fresh engineering pass outs remains unemployed even after having degrees and diplomas in various specialized technical fields and also after spending hefty amounts as fees.

Eleventh Five Year Plan (2007-12) addressed the challenge of making growth both faster and more inclusive. The rapid growth achieved in the recent years demonstrates that India has learnt how to bring about growth, but yet to achieve distributive justice. Government of India has recently (October 2011) announced National Manufacturing Policy (NMP) that envisages to enhance global competitiveness, domestic value addition, technological depth and environmentally sustainable growth for the manufacturing sector. NMP recognizes that in future manufacturing will be the growth engine of Indian economy. It also envisages additional employment generation to the tune of 100 million persons in the manufacturing sector by 2022.

In this backdrop, it becomes imperative to retain the technical manpower on its tracks rather than diverting to other non professional areas just due to employability and survival issues. This paper explores the issues related to the importance of industry-academia linkages and the need for setting up finishing schools or modular courses in the engineering and polytechnics with the active participation of industry. This approach would effectively address the skill shortage issues and can find a meaningful solution to employability of fresh engineering graduates and also for providing livelihood options. The paper has been divided into five sections. Section II discusses general human development parameters including labour productivity across various states during the post independence era. Section III discusses the issue of employability of fresh engineering graduates Section IV discusses India's rank among major countries in the world in terms of important scientific parameters. Section V discusses the remedial measures to address the issue of employability effectively.

SECTION II

Human Development and Labour Productivity in India

Educational Attainment

It may be noted that India was quite backward in terms of various important educational parameters such as literacy rate, Gross Enrolment Ratio etc. Even after six decades of development planning, literacy rates among the states show wide variations. It may be noted that though substantial improvement has been reported during the last six decades, the literacy rates range between a high range of 90.86 % in Kerala to a low of 47.00 % in Bihar during 2010 **(table 1).**

Table 1: Literacy rates in India – Major States: 1951-2001

Sl.	State	Years					
		1951	*1961*	*1971*	*1981*	*1991*	*2001*
1	Kerala	47.18	55.08	69.75	78.85	89.81	90.86
2	Delhi	–	61.95	65.08	71.94	75.29	81.67
3	UP	12.02	20.87	23.99	32.65	40.71	56.27
4	WB	24.61	34.46	38.86	48.65	57.70	68.64
5	MP	13.16	21.41	27.27	38.63	44.67	63.74
6	AP	–	21.19	24.57	35.66	44.08	60.47
7	TN	–	36.39	45.40	54.39	62.66	73.45
8	Maharashtra	27.91	35.08	45.77	57.24	64.87	76.88
9	Rajasthan	8.50	18.12	22.57	30.11	38.55	60.41
10	Bihar	13.49	21.95	23.17	32.32	37.49	47.00

Source: *Economic Survey 2010-11*

In the case of Gross Enrolment Ratio (I-VIII standards) it may be noted that some states such as Kerala, Tamil Nadu and Madhya Pradesh achieved 100 percent while there are states such as Bihar where it is as low as 46.2 percent (**table 2**). Similarly the number of professional colleges and polytechnic also has been found to be quite high in the southern states whereas populous north Indian states lag behind.

Some of the important Human development parameters for India are given in **table 3**. It may be seen that last sixty years have witnessed substantial improvements in all the factors under consideration. Literacy rates in India improved substantially from a very low level of 18.33 % of the population in 1951 to 64.84 % of population by 2001. In the case of infant mortality rate it declined from 115 per thousand births in 1961 to 53 deaths per thousand births by 2008. Life expectancy at birth improved from 37.91 years in 1950 to 66.46 years by 2010. Mean years of schooling (years 25+) substantially increased in India from 0.919 years in 1951 to 4.405 years by 2010.

Table 2: Gross Enrolment ratio, Professional Colleges and Polytechnics

Sl.	States	Gross Enrolment Ratio(I-VIII class or 6-13 years)	Professional Colleges		Polytechnics (Nos.)		Population (2001) (Thousands)
			(Nos.)	*(Per Million Population)*	*(Nos.)*	*(Per Million Population)*	
1	Kerala	100.1	244	7.66	59	1.85	31841
2	Delhi	99.7	40	2.89	16	1.16	13851
3	UP	67.8	335	2.02	110	0.66	166198
4	WB	71.2	192	2.39	48	0.60	80176
5	MP	100.0	287	4.76	45	0.75	60348
6	AP	77.3	1222	16.03	146	1.92	76210
7	TN	112.7	630	10.10	261	4.18	62406
8	Maharashtra	86.8	748	7.72	178	1.84	96879
9	Rajasthan	81.4	261	4.62	29	0.51	56507
10	Bihar	46.2	61	0.73	13	0.16	82999

Source: *Economic Survey 2010-11*

Table 3: Human Development Parameters - All India

Indicator	Years							
	1951	*1961*	*1971*	*1981*	*1991*	*2001*	*2008*	*2010*
Literacy Rates	18.33	28.30	34.45	43.57	52.21	64.84	-	-
Infant Mortality Rates	-	115	-	-	-	-	53	-
Life Expectancy at birth	37.91	44.09	50.77	56.20	58.98	62.5	-	66.46
Mean Years of Schooling (Years 25 +)*	0.919	0.941	1.228	1.216	2.95	3.58	-	4.41
Human Development Index*	-	-	-	0.320	0.389	0.440	-	0.519

Note: ** data for 1981, 1991, 2001 refers to 1980, 1990, 2000 respectively*

Source: *1. Economic Survey 2010-11*

2. World Population Prospects, April 2011

3. Robert J. Barro & Jong-Wha Lee, April 2010

Table 4 reports Labour Productivity in India during the period 1970 to 2008. It may be noted that Labour Productivity consistently improved during 1970 to 2008 period. GDP growth also exhibited similar trends during the same period. These two indicators reflect the real growth trends in the economy.

Table 4: Labour Productivity (LP) in India: 1970 to 2008 (at 2000 Constant prices)

Years	US $ '000 per Worker	LP Growth %	GDP (Rs.Crores)	GDP Growth Rate %
1970	0.462		473591	
1975	0.488	1.10	550072	3.04
1980	0.521	1.32	641921	3.14
1985	0.559	1.42	815049	4.89
1990	0.627	2.32	1083572	5.86
1995	0.703	2.31	1396974	5.21
2000	0.834	3.48	1864301	5.94
2005	1.000	3.70	2615651	7.01
2008	1.176	5.55	3345712	8.55

Note: *GDP & GDP Growth authors own estimation*

Source: *APO Productivity Databook 2011, Asian Productivity Organization*

Table 5 provides comparative productivity estimates per worker across major Asian countries and USA (World Competitiveness Yearbook 2011). It may be noted that labour productivity levels in large economies such as China Mainland and India are quite low as compared to other developed economies. Very low levels of labour productivity in India and China indicate that considerable scope exists for improving the labour productivity in these countries.

Table 5: Comparison of Overall Productivity in 2010 (GDP per person employed)

Country	US $	Rank	Growth %
Japan	87277	17	4.37
India	3473	59	5.97
China Mainland	6391	57	8.39
ROC (Taiwan)	40998	31	8.56
Hongkong	63941	26	5.84
South Korea	42597	30	4.67
UAE	66389	25	-2.98
Quatar	99729	10	15.35
Singapore	71702	23	10.20
Malaysia	21971	44	5.75
Philippines	5235	58	5.73
USA	105422	7	3.46

Source: *World Competitiveness Year book 2011*

Table 6 provides the human resource requirements in the vocational stream across various sectors in India. It has been estimated that there is an annual requirement of 8 million persons in various sectors of the economy. The highest requirement has been estimated for the building construction and real estate sector with 2.36 million persons followed by Auto and Auto Components sector with 1.35 million.

Table 6: Incremental Human Resource requirement in Vocational Stream

Sl.	Sector	Annual Requirement in Vocational stream (in '000s)
1	Textiles (Spinning, Fabric Processing, Garmenting)	1020
2	Electronics & IT Hardware	81
3	Leather	282
4	Organized Retail	991

5	Gems & Jewellery (Including jewellery retail)	247
6	Building, Construction & Real Estate	2365
7	Banking, Financial Services & Insurance	197
8	Furniture & Furnishings	195
9	Auto & Auto Components	1350
10	Tourism & Hospitality	169
11	Food Processing	530
12	Construction Materials & Building Hardware	39
13	Chemicals & Pharmaceuticals	34
14	Transportation, logistics & Warehousing	506

Source: *National Skill Development Council Report*

SECTION III

Employability of Fresh Engineering Graduates

Recent years have reported considerable unemployment in the technical job market. A large number of fresh engineering graduates could not obtain campus placements or employment opportunities in the labour market. This scenario hitherto was unheard of. If we fathom the unemployment issues further it becomes explicit that a number of factors are responsible for the present situation.

It may be noted that the years succeeding economic liberalization in India witnessed a large number of professional technical colleges opening in the private sector as a result of the opening up of the education sector. Economy also witnessed unprecedented demand for technical manpower for creating the basic infrastructure for the growing economy and also the growing manufacturing sector. Many of these technical institutions started inducting students in engineering courses after taking capitation fee or building construction loan from the students even before setting up the basic teaching infrastructure and also without recruiting qualified teaching staff. However, these institutions kept on churning out fresh engineering graduates on a regular basis resulting in a steady flow of low skilled technical manpower that the industry found unemployable. This issue has national proportions since large quantities of resources are being spent on producing an engineer but the manpower was wanting in so many terms for employment in the industry. This situation is leading to frustration among the fresh engineering pass outs and they are migrating to non technical areas for earning a livelihood; thus the country started losing the services of the technical manpower.

Eleventh Five Year Plan (2007-12) addressed the challenge of making growth both faster and more inclusive. The rapid growth achieved in the recent years demonstrates that India has learnt how to bring about growth, but yet to achieve distributive justice. Government of India has recently (October 2011) announced National Manufacturing Policy (NMP) that envisages to enhance global competitiveness, domestic value addition, technological depth and environmentally sustainable growth for the manufacturing sector. NMP recognizes that in future manufacturing will be the growth engine of Indian economy.

It also envisages additional employment generation to the tune of 100 million persons in the manufacturing sector by 2022. Moreover, India is also gearing up to become the manufacturing hub of the world. In such a scenario, the technical manpower needs to be retained in their tracks for the development of the manufacturing sector in India.

Another important factor is that there is negligible interaction between industry and academic institutions in India resulting in the technical institutions churning out generalists rather than specialists who the industry can directly employ. There is also considerable lack of awareness regarding the importance of finishing schools for the fresh graduates so that they can be found employable by the industry.

A recent study undertaken by the NTMIS Nodal Centre of Cochin University of Science & Technology reveals that the unemployment rate among the 2008 batch of passed out engineering students is 30.5% (**Table 7**). In the case of MBA unemployment is 47.8% and for pharmacy students it is 22.4 %. Among the employed engineering students 51.8% of them migrated to other states for employment.

Table 7: Details regarding employment of 2008 batch of CUSAT students

Sl.	Particulars	Number			
		Engineering	MBA	Pharmacy	Hotel Mgmt.
1	Percentage of employment for the batch passed in 2008	53.9	49.8	49.3	81.5
2	Percentage of unemployment for the batch passed in 2008	30.5	47.8	22.4	0
3	Percentage of higher studies for the batch passed in 2008	13.8	2.4	26.4	18.5
4	Percentage of self-employment for the batch passed in 2008	0.06	0	0	0
5	Migration for employment to other states for the batch passed in 2008	51.8	17.5	5	34
6	Migration for employment to other countries for the batch passed in 2008	7.8	12.5	6	6.8
7	Average emoluments for the batch passed in 2008	16160	13927	5815	7227

Source: *NTMIS-Nodal Centre, CUSAT*

SECTION IV

R&D and Scientific Research – A Comparative Analysis

Apart from the employment scenario, the lack of industry-academia interaction results in depleted research and lack of innovations and patents produced at Indian technical institutions as compared to other developed countries. Many well known academic institutions in the west closely interact with the industry especially in research areas resulting in path breaking inventions and innovations in the production process. As a result of the lack of inter-linkages between industry and academia, India lags behind other major countries in terms of R&D expenditure, R&D personnel, number of patents awarded, scientific articles, knowledge transfer, innovative capacity etc.

Quality of technical manpower and scientific research in any country is based on the cumulative result of continued expenditure on R&D over the years. It may be seen from **table 8** that USA leads expenditure on R&D with 2.77% of GDP whereas India spends only 0.80% of GDP on R&D. In the case of India, the share of Business Expenditure on R&D is only 0.24% of GDP wh .eas Government is the major contributor of R&D expenditure with 0.56% of GDP. In all other developed countries and China the share of Business Expenditure on R&D as percentage of GDP is higher than Government share.

The data indicates that there is an urgent need for business establishments in India to focus more on R&D to become globally competitive. This lack of spending in R&D gets reflected in poor scores with respect to science degrees, scientific articles and number of patents received by India as compared to other leading countries in the world.

Table 8: Scientific Research in India – Comparative analysis of Major Countries (2009)

Country	*Total Expenditure on R&D (% of GDP) 2009*	*Business Expenditure on R&D (% of GDP) 2009*	*Total R&D Personnel Nationwide ('000s) 2009*	*Science Degrees (% of total first University Degrees in Science and Engineering)*	*Scientific Articles published by origin of author 2007*	*Number of Patents granted to Residents 2009*
				2006		
USA	2.77	2.01	–	31.86	209695	79803
Japan	3.45	2.70	882.7	62.73	52896	153755
UK	1.87	1.16	330.3	35.02	47121	2082
China	1.70	1.25	2291.0	54.93	56806	47975

Russia	1.23	0.77	848.7	22.50	13953	22328
Brazil	1.23	0.58	397.7	16.46	11885	234
India	0.80	0.24	391.1	23.47	18194	1385
Malaysia	0.84	0.51	15.2	46.60	808	269
Philippines	0.10	0.05	17.1	24.73	195	16

***Source:** IMD World Competitiveness Yearbook 2011*

SECTION V

Policy measures required to address the issue of employability

Government of India has recently established National Skill Development Corporation (NSDC) for providing momentum for participation by private players in skill development and also to address the issue of skill shortages at various sectors. The Cabinet Committee has already made advance projections of the requirement of skilled manpower till 2022. It has been reported that in the existing training institutes and technical institutes facilities are awfully short of supplying the required number of technical manpower. Under the skill development programme it has been suggested to start short term modular courses for job oriented professions to hone the skills of the technical manpower.

A similar approach need to be taken up for the fresh engineering pass-outs also with a view to utilize the services of this vast pool of technical manpower effectively in the industry. In this background there is an urgent need to start finishing schools at the engineering and polytechnics with a view to make the fresh engineering pass outs employable by the industry.

For making the system of finishing schools a workable proposition, it is necessary to get the involvement of Industry Associations and leading manufacturers as partners in this activity through a public private partnership mode. Educational institutions need to take a lead in this effort with a view to make their wards employable and also as the architects of the vibrant emerging economic super power India.

Government of India has recently chalked out a National Manufacturing Policy to create additional 100 million employments in the manufacturing sector in the coming ten years. This opens up wide avenues for the fresh engineering students who are graduating in the coming years. India also aims to become the manufacturing hub of the world along with China. All these developments are creating the right setting for the future engineers and diploma holders in technical areas to become partners in the manufacturing activities and they could get employment and also decent livelihood as manufacturing has been identified as growth engine of future development of India.

References

1. APO Productivity Data book 2011, Asian Productivity Organisation, Tokyo, Japan
2. CUSAT, Employment Scenario of Technical Manpower (Degree Holders) Kerala - 2008, National Technical Manpower Information System, Nodal Centre for Kerala, Kochi
3. Economic Survey 2010-11, Ministry of Finance, Government of India, February 2011, Oxford University Press, New Delhi
4. Human Development Report 2011 – Sustainability and Equity: A Better Future for All, UNDP, New York, USA
5. IMD World Competitiveness Yearbook 2011, Institute for Management Development, Lausanne, Switzerland
6. India- Human Development Report - Towards Social Inclusiveness 2011, IAMR, Planning Commission, Government of India, Oxford University Press
7. Jungsoo Park, ADB Economics: Working Paper Series, No. 227, Asian Development Bank, October 2010
8. Robert J. Barro & Jong-Wha Lee, A New Data Set of Educational Attainment in The World, 1950–2010, Working Paper 15902, NBER, Cambridge, April 2010

6

Employability of Engineering Graduates in India in the Era of Globalisation

Problems and Prospects

*V. Upadhyay and Vandita Sahay**

ABSTRACT

India produces a huge number of engineering graduates every year but most of them are unemployable due to lack of proper skill and knowledge. The market demand for young engineers in the country in terms of quality remains largely unmet because most graduates who come for job interviews fall far short of required quality standards. This indicates a lack of quality orientation of engineering education being imparted in the country. Most of the growth in engineering education during the last couple of decades has been in the private sector although during the last few years central government has taken major initiatives in this area by opening several new IITs and NITs spread across several states in the country. However, one major constraint facing both these

* V. Upadhyay, Professor of Economics, Department of Humanities and Social Science, IIT Delhi and Vandita Sahay, Research Associate, DMS, IIT Delhi.

private and public engineering institutions is the acute shortage of quality faculty. Our paper discusses problems of quality of engineering graduates, their skill development and related issues. We also look at the sectoral composition of job offerings to IIT Delhi graduates and postgraduate students through campus placements in recent years.

Introduction

In the age of globalisation, India has the potential to be a global technology leader. The Indian industry is competing globally in software and even in areas such as automobiles, chemicals, pharmaceuticals and engineering equipment. A critical issue that arises, for the future success of Indian industry, is related with the growth and quality of engineering education in India.

Since independence, the government's education policy's initial focus has been to provide adequate number of engineers that are required for the economy. The setting up of the Indian Institutes of Technology (IITs), the Regional Engineering colleges (and their subsequent conversion to the National Institutes of Technology) were targeted at achieving this. Engineering in India is a preferred option at the 10+2 level. This has resulted in a surge in opening engineering colleges primarily in the private sector. Engineering education has expanded remarkably as indicated by rapidly increasing number of engineering institutions, massive increase in the students' admissions and increase in number of engineering disciplines.

Despite this, industry leaders complain about the absence of quality engineers in their industry. This is accompanied by significant unemployment rates amongst graduating engineers. Insufficient supply of quality skills in the freshly passed out graduates is one of the main impediments to further economic growth in India. The Indian economy grew more than 8% on average over the past 5 years, including the year of the unprecedented financial crisis in 2009. However, the skill shortage is still one of the major constraints in most industries in India (World Bank, 2009).

According to Mr. N. R. Narayana Murthy, a major requirement for creating a world class educational institution is freedom of action and choice (Murthy 2007). According to him, there are four critical requirements of a world class institution- leadership, mindset, meritocracy, and humility. The major difference that Mr. Murthy has observed between the institutions abroad (like Cornell, Wharton and Stanford) and in India is that, in the case of the former, the leaders of the institutions are the ultimate decision-makers on what needs to be done to further the cause of the university, whereas in India, we still have not graduated to a level where we can take our own decisions.

With increase of quantity there should be monitoring for quality also. The development of higher education in thelast six decades presents a mixed picture. With quantitative success there is qualitative failure. Massive production of engineers and their lack of employability, reveals a low standard of education. In October 2007 union HRD minister commented about higher education as a sick child and union minister of state for HRD expressed disappointment about the quality of higher education.

A 60-page World Bank paper titled 'Employability and Skill Set of Newly Graduated Engineers in India' (2011) by Andreas Blom and Hiroshi Saeki reveals that "64 per cent of employers are only somewhat satisfied or worse with the current engineering graduates in the country." Under the Technical Education Quality Improvement Programme, the World Bank has aided several engineering colleges in different states, including Karnataka, to improve the quality of education and to make engineers more employable. The paper points out that the skill sets of fresh engineers are inadequate. "Indian employers are less satisfied with their engineers compared to U.S. employers." Though the Indian higher education system has massively expanded production of engineers, there has been a significant decline in the quality of students. According to the authors, employers pointed out those fresh engineering graduates were most lacking in high-order thinking skills, such as critical thinking and problem solving. Besides, employers said that the graduates lacked soft skills, integrity, teamwork, reliability, and willingness to learn.

The question that arises is whether the quality of engineers produced by India is up to the mark? Are the numbers of engineer graduates really skilful to handle market? Do all the graduates get employment? Does there exist a quality and quantity mismatch? And also a demand and supply mismatch?

Quality of Engineers

The US National Science Foundation (NSF) Task Force defines Quality Engineering Education as follows:

"Quality Engineering Education is the development of intellectual skills and knowledge that will equip graduates to contribute to society through productive and satisfying engineering careers as innovators, decision makers and leaders in the global economy of the twenty first century."

Quality Engineering Education demands a process of continuous improvement and innovation in student, employer and societal satisfaction by jointly evaluating and refining the system, practices and culture of engineering education institutions (Nataraja 2000).

Quality in education can be defined with the following parameters (Veeranna.D.K et al 2011)-

- Excellence in education by developing a questioning mind,
- Value addition in education and encouraging creativity,
- Conformance of education output to planned goals, specifications and requirements, and
- Adaptability and the eagerness to learn new things.

According to a study 'Engineering Education in India,' by Rangan Banerjee and Vinayak P. Muley, 2008, it was estimated that 3.5 lakh engineering degrees, 23,000 engineering Masters degrees and about 1,000 PhDs were awarded in India. India's doctorate degrees are less than 1% of graduate engineering degrees. The percentages of doctorate degrees to engineering degrees is much higher for most of the other countries studied (9% USA, 10% UK, 8% Germany, 3% Korea). The study observes that one of the biggest advantages of the top engineering colleges in India is that their admission process is highly selective- approximately 2-3% of the applicants are selected through a Joint Engineering Exam (JEE). This is much stringent compared to that for reputed international universities.

But sadly, the Indian engineering education system has been unable to attract the best engineering students towards post-graduate studies. The study finds that the IITs and Indian Institute of Science (IISc) contribute to less than 1% of the engineering graduates in the country, 20% of the M.Techs and 40% of the Ph.Ds. Only about 1% (or less) of the graduating B.Tech class of an IIT opt for an M.Tech in India, while only 2% of the graduating M.Tech class opt for Ph.D in India. This shows that very few engineering graduates opt for higher education i.e. they are not interested in specialising in their respective fields. About 75% of the engineering graduates are product of private engineering colleges.

Employability of graduates

As verified by various surveys, more than 70 percent of our engineering graduates are not employable. Dr Abdul Kalam has rightly said that India does not have problem of unemployment but "unemployability."

Engineering colleges in India now have seats for 1.5 million students, nearly four times the 390,000 available in 2000, according to the National Association of Software and Services Companies, a trade group. But the fact remains that 75% of technical graduates and more than 85% of general graduates are unemployable by India's high-growth global industries, including information technology and call centres, according to results from assessment tests administered by the group.

The challenge to find suitable employment for the youth without compromising on quality is especially pressing given the country's more youthful population than the U.S., Europe and China. More than half of India's population is under the age of 25, and one million people a month are expected to seek to join the labor force here over the next decade, the Indian government estimates. The fear is that if these young people aren't trained properly with adequate skills and knowledge, they will be unable to participate in the country's economic growth.

Are the Indian Graduates really Unemployable?

There exists a high level of dissatisfaction which is corroborated by other studies on skills of the Indian engineering graduates. For instance, the NASSCOM and McKinsey report (2005) finds that most of engineering graduates are not employable by multinational companies. The report predicted that the Indian industry would face huge talent shortage in the coming years. It brought out the fact that currently, only 25% of fresh engineers and a mere 10% of fresh general graduates are actually employable (also see Gopalakrishnan, 2011 and the WB report cited above).

The educational and skill profile of existing workforce in India is very poor and is primarily responsible for its low productivity. Though enrolment in academic institutions are high in numbers; more than 90% in primary classes, around 60 % in upper primary classes, more than 30 % in higher secondary and above 10 % in higher education, the percentage of people having marketable skills is extremely low (Agarwal, 2007).

Keeping pace with the globalised economy's demand with shifting focus on knowledge-workers and skilled manpower driven employment structure, India's youth needs to be trained and empowered with such a value-adding education, which inculcates the necessary 'employment skills.'

Shortage of Faculty

The country is facing the problem of shortage of teachers for many years. Currently, the higher educational institutions are facing the biggest challenge of acute shortage of qualified and competent faculties. According to an article "Where have engineering teachers gone?" in The Hindu last year, the country faced a shortage of more than 3, 00,000 teachers in its institutions of higher learning. In engineering education alone, the shortage is more than 1, 50,000. Also, the teacher-student ratio is badly skewed. Overall, it is nearly 21 students for each teacher whereas the University Grant Commission (UGC) norm for the undergraduate courses is 1:15 and for the postgraduate courses it is 1:13.5.

From 30 per cent to 50 per cent of the teaching posts are lying vacant in the country's top engineering institutions. The situation is even worse in hundreds of private self-financing colleges. The All India Council for Technical Education (AICTE) currently stipulates a 1:14 teacher-student ratio for engineering institutions. Plans are going on to bring down the teacher-student ratio in engineering institutions to 1:12. The ideal teacher-student ratio, according to experts is 1:10, which is the international standard.

The Central engineering institutions, including the NITs and the IITs, started facing the faculty shortage with a 27 per cent increase in number of seats for the other backward classes (OBCs) over three years from 2008 to 2010. This increase resulted in a cumulative increase of 54 per cent in student strength. Faculty requirement has also increased by a similar proportion which largely remains unfilled for various reasons. The lack of recruitment of supporting or non-teaching staff in recent years has compounded the problem.

The shortage of faculty affects the students in many ways. When there are not enough teachers, there is a lack of attention from teachers resulting in an eventual dilution in quality. Besides, the choices before the students for their elective subjects are considerably reduced.

High teaching load has effect on teachers' research output and doctoral and post-doctoral research at these institutions, which deprives India of any single technical institution of very high international repute.

The government has in recent years set up several new IITs and NITs spread across various states in the country. The infrastructural development at these institutions is however very slow. It will take several more years for these institutes to develop proper infrastructure (buildings, labs, etc). However, even more serious issue facing these institutes is to find adequate number of faculty members who meet required qualification criteria.

Metro-man Mr. **E. Sreedharan** *finds the quality of a majority of engineers in the country below acceptable standards. He blames this on the uncontrolled growth of private engineering colleges, with no regulatory body in sight. He says that "nobody is controlling quality – mass-producing engineers without any quality is of no use to us. There should be a body to regulate and fix the minimum standard required for engineering education and bring about uniformity in the whole country."*

Changing Job Opportunities for Engineering Graduate: A Case Study of IIT Delhi

Engineering graduates passing out from IITs do not as such face the problem of unemployment as almost all of those who do not go for higher studies are able to find good jobs. There is however a continuous change in sectoral composition of job offerings depending on changing economic condition in the economy. To analyse the recent trends in this respect, in the absence of complete employment information, we look at the data for number of engineer graduates who were placed from IIT Delhi for three years- 2008-09, 2009-10 and 2010-11 given below.

2008-09

Track	BTech	Dual	MTech	Other
Analytics	15	10	3	0
Consulting	53	26	31	0
Core (Technical)	105	13	100	8
Finance	36	18	0	0
Information Technology	46	18	30	0

Management	0	0	0	0
Other	72	12	86	13
PPO / Departmental Effort	3	2	•3	0
Total	330	99	253	21

2009-10

Track	BTech	Dual	MTech	Other
Analytics	20	11	5	5
Consulting	39	11	13	0
Core (Technical)	70	15	80	1
Finance	59	23	2	1
Information Technology	30	23	31	4
Management	1	0	0	0
Other	61	13	50	2
Teaching & Research	8	1	23	17
PPO / Departmental Effort	32	12	24	4
Total	320	109	228	34

2010-11

Track	BTech	Dual	MTech	Other
Analytics	17	4	3	3
Consulting	70	22	9	0
Core (Technical)	81	11	79	5
Finance	40	27	1	0
Information Technology	46	28	57	9
Management	7	1	0	0
Other	54	14	70	3
Teaching & Research	2	1	22	4
PPO / Departmental Effort	17	1	37	0
Total	334	109	278	24

The above table shows the number of campus recruitments for B. Tech, Dual, M. Tech and Other courses students. The data show that in 2008-09 maximum number of B. Tech students, i.e 105, were hired for core (technical) jobs, which reduced over the years and came down to 81 in 2010-11. While, the number of graduates recruited by the IT, finance and consulting sectors has seen fluctuations over the years. Very few graduates are recruited in teaching and research sector. This is mainly because of lack of interest on the part of graduates in these areas.

Compared to B. Tech students, more M.Tech students opt for core (technical) jobs. However, there appears a gradual shift from core (technical) jobs to IT sector jobs even for M.Tech students.

Conclusion and Recommendations

If engineering institutions want to improve the employability of their graduates, they have to focus on reducing important skill gaps through improvements in curriculum and teaching methods. Curriculum up gradation from time to time is very important in this respect. Higher weightage needs to be given for assignments to students which encourage independent thinking and application of tools on real life problems. Industrial training by students need to be emphasised and be made suitable in terms of its effectiveness to assure that they derive clear benefits in terms of better job opportunities later on. Only through such changes in the teaching-learning process will the future engineers become more employable.

References

1. Agarwal, P. (2007). Higher Education in India: Growth, Concerns and Change Agenda, *Higher Education Quarterly*, 61: 197–207.
2. Gopalakrishnan, A.(2011). Employability and transition to labour market in emerging economies: An Indian perspective on Higher education, presentation under the theme *Seeding and harvesting. Students in mass higher education-Track 4* at the EAIR 33rd Annual Forum in Warsaw, Poland held on 28-31 August 2011.
3. Murthy, N.R. (2007) "Building World Class Educational Institutions in India: Challenges and Prospects", Vikalpa, Vol 32, No. 1, January - March 2007.
4. NASSCOM-McKinsey (2005). Report on Extending India's Leadership of the Global IT and BPO Industries.

5. Nataraja, R. (2000) *"The Role of Accreditation in Promoting Quality Assurance of Technical Education"*, International Journal of Engineering Education, Vol. 16, No. 2, pp. 85-96.

6. Veeranna, D.K., K. Joshi, Anand and.G.N., Sahadeva (2011) *"Quality Issues In Engineering Education: An Indian Perspective"*, International Journal of Engineering Science and Technology, Vol. 3 No. 3 March 2011.

7. World Bank, 2009. India's Investment Climate, Voice of Indian Business, World Bank, Washington DC, USA.

7

Labour Market Situation of Agricultural Manpower

Some Issues

*I.C. Awasthi and Indra Kumar**

EXECUTIVE SUMMARY

This paper examines the labour market situation of agricultural manpower, particularly of agricultural engineering manpower that has profound implications to agriculture growth and food security.

Education and skill are critical for this sector as a substantive proportion of our workforce in agriculture are either illiterate or having schooling up to primary level. In the fast changing economic milieu and WTO regime, a more serious challenge is how to improve skills and technical education. High level manpower is the most critical factor for research and development and for enhancing inventive capacity. This paper focuses on the serious concern about the quantitative and qualitative aspects of human capital, which suffers from mismatches on both demand

* Head, (NTMIS) and Asstt. Director, respectively at the Institute of Applied Manpower Research (IAMR), Delhi. Authors are thankful to Mr Yogesh Kumar, Joint Director, IAMR for his useful comments and suggestions.

and supply counts. Almost one third of the agricultural graduates are unemployed thereby raising concerns about their quality and pitifully low wage premia associated with educational attainments. Our enquiry has revealed this phenomenon clearly.

Disparity of institutions is conspicuous and this has led to serious imbalances across states and regions. This Paper argues that how a second green revolution can be launched without a strong agricultural human capital base that is critical and so central in the fast changing globalised regime. How can a relatively faster economic growth be sustainable without robust agricultural growth? A huge agricultural education infrastructure has been created since early planning period yet it is still a puzzle that it has no substantively visible impacts in agriculture growth. This Paper, therefore argues for the urgent revamping of agriculture education and research policy.

Introduction

It has been argued that any structural transformation is considered success-ful when it is accompanied by corresponding changes in the struc-ture of work force. In spite of perceptible changes in the share of gross domestic product (GDP), there has not been any substantial change in the structure of work force over the years (Chakravarty, 1988). For instance, the proportion of the workforce dependent on agriculture has declined from about 74 per cent in 1972–73 to about 57 per cent in 2004–05, while the share of national income originating from agriculture has declined from about 45 per cent to about 18 per cent during the same period. This kind of structural change has resulted in low productivity in the agriculture sector.

Agriculture continues to be the engine of inclusive and accelerated economic growth and livelihood security. Faster growth of agriculture is a pre-requisite not only to achieve self-reliance of food security at the national and at household levels but also to reduce poverty and regional inequity.

While there has been tremendous progress in the field of agriculture, the pattern of growth has brought about serious problems of regional disparity. Also, there are numerous problems associated with agriculture both from demand side and supply sides. Some of the demand side bottlenecks are controls on movement, storage and sale of agricultural products, unfavourable price regime, low value addition, distortions in the incentive regime, undue price fluctuations in world markets, and in the World Trade Organisation (WTO) regime. The supply side constraints essentially relate to capital inadequacy in terms of decline in the public sector investment, lack of infrastructural support, slowing of net irrigation expansion, falling growth in total factor productivity, bottlenecks in institutional reforms, etc. (Bhalla, 2006; Rao,2006).

Agriculture growth has sharply decelerated in recent years leading to serious repercussions in the overall growth of the economy. Agriculture had registered reasonably respectable growth during 2005–06 to 2007–08 (5.2, 3.7, and 4.7 per cent, respectively) along with an overall 9 per cent growth during this period. Subsequently, agriculture growth dipped to 1.6 per cent in 2008–09 and further declined into the negative (-0.2 per cent) in 2009–10 (Economic Survey, 2010).

Slower agriculture growth has been the principal concern for overall growth of the economy and National Development Council (NDC) in its meeting (29th May, 2007) decided to launch a special additional central assistance scheme in order to reorient and rejuvenate agriculture development strategies during Eleventh Plan period. NDC reaffirmed to achieve 4 per cent agriculture growth per annum, which is the essential element for broad based and more inclusive growth. There are serious apprehensions of attaining 4 per cent growth in the light of aggregate domestic demand (Vaidyanathan, 2010). In most of the reasoned discussions, capacity building is often either missed out or not examined seriously. Though we have created a large agricultural capacity building infrastructure that includes agricultural universities, agricultural colleges, ICAR institutions, Krishi Vigyan Kendras, and special missions etc., this remains one of the critical inputs that has profound implication to agriculture growth.

This short note attempts to look into some of the aspects of human capital base on the nation wide data collected through National Technical Manpower Information System (NTMIS). However, this note has been written within certain limitations. First, the paper has a somewhat limited scope of discussion as it primarily discusses few disciplines under agriculture and allied activities. Second, data is not adequate to study the problem in its entirety, as it covers only AICT institutions. Almost half of the states/UTs do not report any information therefore data is not available to make any robust and time series analysis. ***Third, though it is claimed that data is collected through follow up enquiry, institutional and establishment surveys from a nationally representative sample but evidently it lacks such sampling rigor.*** Nevertheless, some clues are discernible from this analysis that might aid policy in order to reorient and restructure education and labour market policies or even help in mid course corrections.

The paper has been arranged in to five sections. The first section briefly analyses the concerns of the agriculture sector. The second section deals with educational challenges in the perspective of new agriculture. The third section analyses the institutional capacity, intake capacity, outturn patterns and shortage of teachers. The fourth section addresses the labour market features in terms of activity status, employment, unemployment, sectors of employment and emoluments, and turnover and migration etc. The last section draws conclusions and analyses the policy imperatives.

Challenges of Skilled Manpower

The educational level of a person acts as a crucial factor in determining his/her employability and earnings. Technical education is a major dimension of human resource development and will play an increasingly important role in the new economic perspectives to produce relevant higher quality skills, which are capable of ensuring higher growth and productivity in a knowledge based society (Jha, 2010). Technical courses need to be qualitatively improved so as to make them marketable, otherwise these would continue to become less relevant to the needs of market. In order to make them relevant there is a need for evolving flexible education and training systems to ensure quality and relevance; ensuring balance between cognitive and manipulative aspects of education and training to improve the employability of skills; establishing strong industry-education/training sector linkages to facilitate continuous and relevant changes in the curriculum and training methods; to reduce mismatch by strengthening mechanisms for continuing education, apprenticeship and on-the-job training and to attach prime importance to teacher training, as well as training of trainers in a dynamic economic environment. This calls for labour market monitoring on a continuous basis in order to ensure optimum utilization of skilled manpower both on the demand side and supply sides. An accreditation council should be established and entrusted with the responsibility of accreditation of all educational institutions, updating them on yearly basis.

If a second green revolution is to be launched, widening the base of human capital in agriculture education and engineering is a prerequisite. This would require not only a qualitative expanding of the base of technical manpower but also improving the educational base of our peasantry to meet the needs of a changing world of agriculture. An overwhelmingly large proportion of our agricultural work force is illiterate or has primary level of education. The National Sample Survey Organisation's (NSSO) findings suggest that in 2004–05, 74 per cent of agricultural labour and 58 per cent farmers were illiterate or have received primary education. Introducing new agriculture in the new millennium would be extremely difficult with such a level of illiteracy. Further, commercialization of agriculture in a highly competitive environment has resulted in many crop enterprises increasingly requiring education as they are skill intensive (Chadha, 2006). This necessitates the urgency and desirability of relevant educational and training activities to face the challenges of new agriculture in the perspective of globalization.

Institutional Capacities

Number of Institutions

There has been a huge expansion of technical institutions over the years and in particular the 1990s saw phenomenal growth in both degree and diploma institutions (Chart-I). Engineering institutions for degree courses increased by 9.1 per cent during 1990 and 1995 and it went up to 14.1 per cent in between 1995 and 2000. Then growth decelerated to 11.5 per cent in 2005 (Table 1). Surprisingly, these institutions saw a sudden decline in absolute numbers in 2008. However, this drastic fall is unconvincing and appears so primarily due to non-reporting of figures as many as 10 major states (namely, Jammu & Kashmir. Himachal Pradesh, Uttar Pradesh, Kerala, Karnataka, Andhra Pradesh, West Bengal, Orissa, Madhya Pradesh and Delhi). Similarly, diploma institutions grew at about 5.1 per cent during 1990 and 1995 then the growth decelerated to 3.2 and 2.4 per cent during the periods between 1995–

2000 and 2000–2005, respectively. Thereafter they recorded negative growth which again is due to non-reporting error. Also, there is large data gap as no data was reported from about 20 states/UTs that makes extremely difficult to make generalisation.

Over time there is a significant imbalance in the geographical spread of technical education both in engineering and diploma institutions. The four southern states and Maharashtra combined have almost 60 percent of the country's engineering institutions. While the eastern and northern states constitute just 16 percent of the total. In terms of undergraduate enrollment, the divide is even more evident (WENR, 2007). This has been corroborated by our nation wide primary data as well.

Southern states (Kerala, Karnataka, Tamil Nadu and Andhra Pradesh) have recorded extraordinary high growth. Almost half of these institutions are concentrated in this region. This is followed by western states (Gujarat, Rajasthan and Maharashtra) which constitutes about one fifth of all the institutions (ignoring aberrations).

The north states (Jammu & Kashmir, Himachal Pradesh, Uttar Pradesh, Punjab and Haryana) constitute little less than one fifth institutions. Eastern states (West Bengal, Orissa) make up less than one tenth and UTs and north eastern states have abysmally low share less than 2 per cent.

Chart I

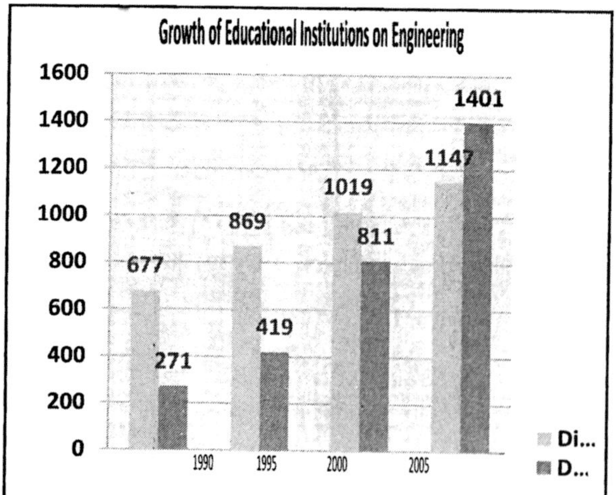

Intake Patterns

Table 2 shows intake pattern of agriculture and allied engineering graduates (agricultural engineering, dairy technology, environmental engineering, food technology, forestry, fishery, marine engineering and sugar technology) by states. The difference between the sanctioned and actual intake is clearly discernible and the actual intake has been consistently lower than the sanctioned intake throughout the period during 2001 to 2008 (Chart II). The states showing declining actual intake are Arunachal Pradesh, Assam, Haryana, Himachal Pradesh, Karnataka, Kerala, Madhya Pradesh, Maharashtra, Orissa, Punjab, West Bengal, Delhi and Chandigarh. The exceptional cases are Tripura with a marginally higher actual intake than the sanctioned and Andhra Pradesh shows perfect correspondence between actual and sanctioned intake through out. In a couple of states (Chhattisgarh, Jammu & Kashmir) actual intake is given but without sanctioned intake. Surprisingly, actual and sanctioned intake has dipped down from 2004 and there has been catastrophic fall from 2006.

How does one interpret this falling intake trend? One could possibly draw some conjectures as why the intake capacity is showing an unrelenting decline over a period of time.

First, agriculture and allied disciplines appear to have saturated in terms of market demand in relation to other emerging disciplines.

Second, there has been lack of correspondence between the supply of outputs from universities/ colleges and requirements of outputs by the industry.

Third, institutions lack or delay in updating course contents and syllabi on continuous basis in view of emerging market trends both nationally and internationally (Singh, 2006; ICAR,2010).

Fourth, there appears to be a lack of collaborative research and inventive capacity in the universities/institutions that severely restricts competing in international markets.

Finally, there is a need to undertake continuous research on the labour market situation with a view to getting signals and to effect necessary changes in the educational system.

It appears to be in contrast that on the one hand there is an urgency to launch a second green *revolution* but on the other the human capital base is awfully low to sustain such a grand endeavor.

Appropriate long term manpower policies need to be formulated in order to actualise the dream of a second green revolution.

Chart II

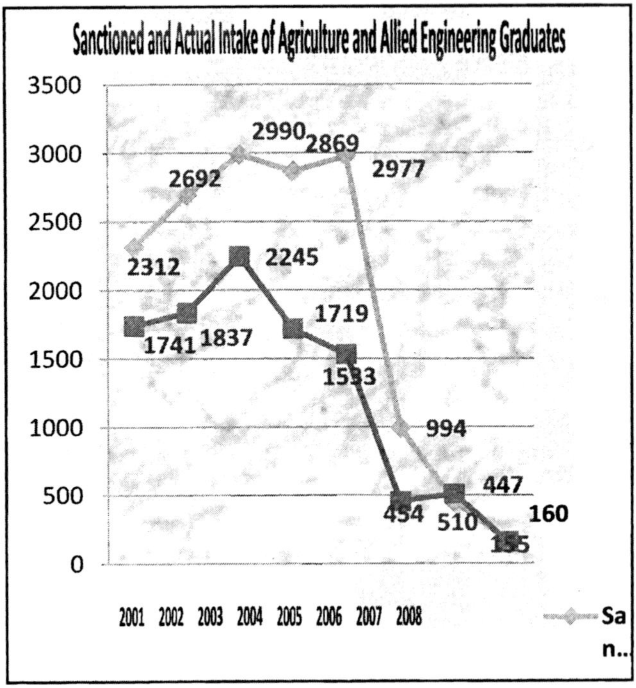

Outturn Patterns

In general, the outturn pattern of all engineering graduates appears to have increased from 2001 to 2005 but then it suddenly dropped down in 2006. Outturn of agriculture and allied graduates at all India level is abysmally low at 0.75 per cent of the total outturn reported in 2008 (Table 3). Even in the state level it turns out to be terribly low. In certain states the outturn of agriculture and allied graduates has declined with an erratic trend (namely, Haryana, Karnataka, Madhya Pradesh) and forms minuscule to total outturns while Andhra Pradesh, Arunachal Pradesh, Kerala, Orissa, Punjab, Tamil Nadu, Uttar Pradesh, West Bengal and Delhi report by and large uptrends, albeit marginally.

Surprisingly, half of the states/UTs did not report outturn and as a result it becomes extremely difficult to make any conclusive generalizations.

The relationship between intake and outturn is clearly discernible in Chart III. Intake grew steadily from 2001 to 2003 then it started continuously declining and thereafter fallen catastrophically during 2005-06. Generally, intake may remain stable in particular year or over the years but outturn has specified gestation period (say, four years) that may vary depending upon the success and drop out rates.

It picked up albeit marginally in 2007 but then declined in 2008. While outturn showed gradual increase until 2005 then there has been dramatic fall in outturn as well. Intake was higher than outturn till 2004 indicating gross mis-utilisation of human resource.

Chart III

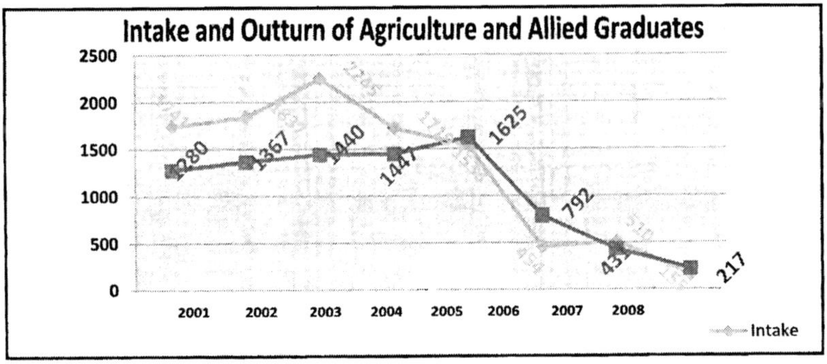

One could possibly surmise that faster growth of institutions resulted in relatively higher intake by the institutions in 1990's but outturn could not be sustained correspondingly. This has led to sheer loss of human capital. However, in 2005 and 2006 outturn outshined the intake, probably due to clearance of backlog of repeaters.

Our postulation, therefore, clearly lend credence that agriculture appears to be at its lowest ebb in the emerging disciplines and there is need for making it relevant to the market needs in order to increase intake and outturn.

Shortage of Teachers

One of the major constraints is the acute shortage of teachers that has hampered education and research in our technical institutions. It has been estimated that there is a shortage of 1.2 lakh faculty in engineering colleges in the country and the quality of existing faculty is a matter of serious concern (Biswas, 2010). In almost every state/UTs, teachers in positions are far below the sanctioned strength and what to talk of adhering to the ratio between teacher and students. This indicates the apathy of our educational system which has had serious implications pertaining to the quality and standard of teaching. Teachers in agricultural engineering form minuscule numbers to the overall strength of teachers in degree/PG courses in engineering and of which substantive numbers are unfilled (Table 4). The shortage of faculty is chronic and results in increased class size, which increases the work load for existing teachers that in turn adversely affects the quality of educational achievements. Lost faculty positions further reduce the capacity of universities to prepare a sufficient supply of qualified teachers.

Institutions are facing a faculty shortage to the extent of 67 per cent all over the country. The U.R. Rao Committee (2003) earlier pointed out that India had a huge shortage of teachers for engineering. The committee's report, *Revitalizing Technical Education (2003)*, describes a technical sector that is expanding at an unsustainable level and is in drastic need of regulation to ensure academic standards are improved (WENR, 2007). Rao Committee report aptly remarked that *"A serious situation has arisen in recent years because of the mushrooming of a large number of private technical institutions and polytechnics. Barring some exceptions, there is scant regard for maintenance of standards"*.

Labour Market Features

Labour market monitoring is a significant process in order to optimize utilization of labour force in the context of changing manpower dynamics (Papola, 1992). It is a continuous assessment of demand for and supply of manpower—in particular skilled manpower—with a view to achieving optimal utilization, when surpluses and shortages of skilled manpower co-exist together. This assessment is necessary both at macro and disaggregated levels for planning and policy purposes. This obviously needs collection of information or the setting up of labour market information system (LMIS) at disaggregated levels. This is followed by analysis of information collected in a systemic manner. Such an analysis in turn helps identifying causes and trends of labour market and provides the signals that lead to culmination of actions either for mid-way corrections or for taking appropriate policy measures.

Activity Status of Engineering Degree Holders

Table 5 sheds light on the activity pattern of engineering graduates for the 2007.The table highlights several findings. First, 29 per cent engineering graduates are unemployed and the proportion of agricultural engineering is still higher at 35 per cent (Chart IV). Such huge unemployment is utter wastage of human capital and our deduction distinctly corroborates the hypothesis that the quality of our graduates is not only poor but not relevant to the market needs. Also it is reported that almost one third of engineers are not employable primarily because of poor quality (Jadhav, 2010). Huge expansion of engineering education without any established accreditation system would continue to become less relevant to the needs of market. Some of the states report even higher rate of unemployment (namely, Punjab 37 per cent, Karnataka 48 per cent, Manipur 45 per cent, Tripura 34 per cent, Maharashtra 32 per cent). This is indicative that the quality of education in our engineering institutions needs urgent attention. It is necessary to revisit our engineering education policy (Table 5).

Second, paid employment is 61 per cent at all India level which is respectable and in some states it is above 75 per cent (namely, Delhi 76 per cent, J&K 89 per cent, Arunachal 89 per cent, Assam 93 per cent, West Bengal 86 per cent, M.P. and 89 per cent). Higher proportion of wage employment is regarded as a positive feature if it is accompanied by reasonably good remuneration that we will look in to subsequent section. Even in the case of agricultural engineering this proportion is lower (53 per cent) than the overall engineering graduates (Chart IV).

Third, overall self-employment constitutes a minuscule proportion (0.36 per cent) and the share of agricultural engineering is higher (2 per cent). Agricultural engineers are probably forced to resort to self-employed activities due to non-availability of paid employment. Low proportion of self-employment is a result of our educational policy and value system that encourages engineers to choose white collar jobs.

Fourth, apprentices also represent small fraction (0.52 per cent) and this reflects poor education-industry linkages and relevance of our outputs to the needs of industry.

Chart IV

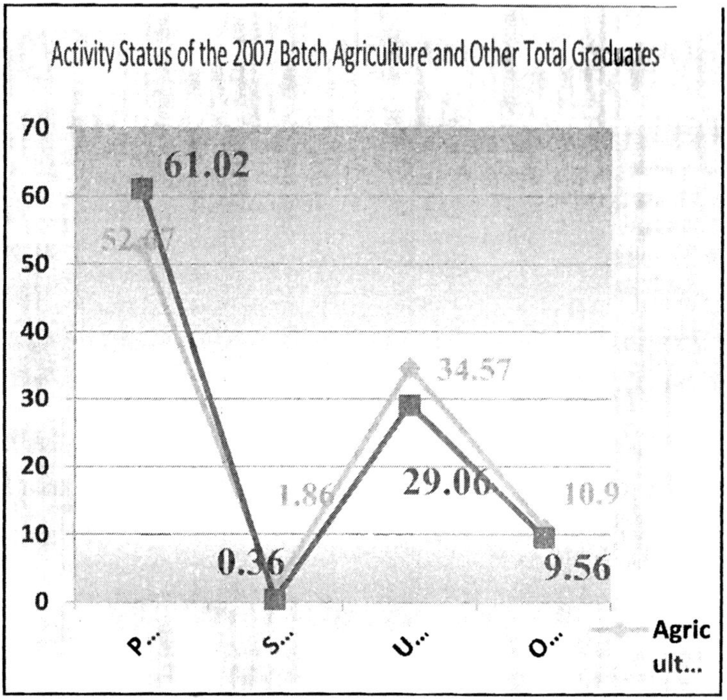

Unemployment

Out of the total number of unemployed graduates reported in 2007, over 99 per cent are looking for jobs and a tiny proportion of them are either not looking for job or are interested in self-employment opportunities. Share of males is disproportionately higher for looking jobs (72 per cent), not looking jobs (55 per cent) and interested in self-employment (86 per cent). The agricultural engineering unemployed form a minuscule proportion of total unemployed (0.31 per cent) and almost all the unemployed graduates seem to be looking for jobs (Table 6). One of the central issues for such a huge unemployment is that no market demand exists for the quality of our engineers produced. This has happened due to mushroom growth of engineering education without setting quality parameters. There is, therefore need to place regulatory mechanism in institutions within private sector to maintain standards of quality.

Sectors of Employment

Private sector is a major employer for engineering graduates which constitutes nearly three fourths (74.5 per cent) followed by public sector units (12 per cent), local bodies (4 per cent), state government (3.3 per cent),central government (2.5 per cent). Agriculture engineering graduates constitutes minuscule proportion (0.24 per cent) to the total engineering graduates by sectors of employment. The highest employment sector for agriculture engineering graduates is the private sector (60.4 per cent), public sector units (12.3 per cent), state government (10.1 per cent), central government (1.7 per cent) and local body (0.9 per cent) in that order (Table 7). It is clear that major source of future employment is going to be the private sector keeping in mind increasing privatisation and government sector will diminish in the provision of employment.

Across disciplines, computer, electronics and communication, mechanical, electrical and Civil engineering are preferred disciplines and textile, agricultural, metallurgy, mining and aeronautical are least preferred disciplines in that order (Table 8).

Private sector is the highest employing sector in all disciplines followed by PSUs except in the case of agricultural engineering. In any case, government (central or state) does not appear to be a major player barring in the case of metallurgy and mining. Local bodies seem to be the emerging segment for employment particularly for computer, electronics & communication and mechanical engineering disciplines. Aeronautical engineering graduates are only being absorbed in private sector and PSUs and similarly mining graduates are employed in the private sector, PSUs, state government and others.

Emoluments

There are no perceptible variations of emoluments between fresher and experienced and that of maximum, minimum and average level of emoluments. But there are large variations between maximum, minimum and average emoluments. Minimum emoluments are almost one third lower than the maximum (Table 9). Also, average emoluments are much lower which is equivalent to emoluments of primary teacher in government schools. The level of emoluments in a way reflects the relative wages and skill premia in the market.

Disciplines in the case of fresher getting maximum emoluments are aeronautical, metallurgy, electronics & communication and computer engineering while lowest paid are agricultural, textile engineering getting less than Rs. 9,000 per month. Average emoluments are highest in the case of aeronautical and followed by textiles, metallurgy and chemical engineering. In the case of experienced, maximum emoluments getting disciplines are aeronautical, electronics & communication, computer and metallurgy engineering and lowest are agricultural and textiles engineering. Average emoluments are highest in the case of aeronautical followed by textile and lowest are for architectural and agricultural engineering (Table 9).

Place of Work

A little over half (52 per cent) of the graduate engineers work within the state, 47 per cent work outside the state and a small proportion of them (1 per cent) work abroad. Males dominate in each of the work places and females have minuscule share.

Industrial/production engineering, architectural engineering, textile engineering and civil engineering have disproportionately higher share of place of work within state, ranging from 65 per cent of civil engineers, 68 per cent textile engineers to 77 per cent industrial/ production engineers and 80 per cent architectural engineers (Table 10).

These disciplines appear to have rigidity in terms of their mobility as some of them are still considered traditional engineering disciplines. Aeronautical engineering, biotechnology, metallurgy engineering, electrical engineering and mining engineering have greater mobility and work substantively out side the state. Some of these are emerging disciplines and have location specificities that demand huge mobility and turnover across the states. Aeronautical, metallurgy, mechanical and civil engineering disciplines registered relatively higher share in work places abroad. This is plausible owing to greater demand of these disciplines world wide.

Migratory Patterns

State wise migratory pattern of engineering graduates for the year 2007 is presented in the table 11. At all India level there is 6.6 per cent of engineering graduates who migrated from other states as per cent of total outturn. Surprisingly, Chandigarh has almost all the engineering graduates from outside the UT. Arunachal Pradesh and Meghalaya have 79 per cent graduate migrants from other states followed by Assam and Himachal Pradesh who constitute of about 38 per cent. Such huge migrants are principally due to expansion of institutions in these states and also in certain cases there is reservation for central government employees. In other states this proportion is significantly lower (less than 10 per cent). Residents of the state who graduated from elsewhere as per cent of total outturn forms a small proportion (less than 1 per cent) with highest share being in the case of Delhi(15 per cent) followed by Himachal Pradesh (9 per cent), Jammu & Kashmir, Kerala and West Bengal (3 per cent).

Net migration as per cent of net outturn appears to be small (5.2 per cent) at all India level but there is wide variations across states. Chandigarh has 50 per cent net migration followed by Arunachal Pradesh (44 per cent), Assam (27 per cent) and Himachal Pradesh (22 per cent). In other states/UTs net out-migration appears to be small varying from less than 1 per cent to about 12 per cent.

Looking from the point of view of courses, overall migrants graduated from other states/UTs as per cent of total outturn is about 6.5 per cent with disparity across disciplines (Table 12). Aeronautical engineering has highest proportion (72 per cent), probably due to limited number of institutions offering degree courses. Metallurgy engineering is yet another course which has relatively higher magnitude of migrants (29 per cent) followed by mining engineering (19 per cent). In other courses the extent of migrants from other courses is low, ranging from 5 per cent in computer engineering to 11 per cent in agricultural engineering. Residents of the state who graduated from elsewhere as per cent of total outturn by courses forms a small proportion (less than 1 per cent) and highest being in the case of metallurgy engineering (6 per cent) and in other courses it is less than 2 per cent.

Net migration as per cent of net outturn forms a small part (5.3 per cent). However, in the case of aeronautical engineering it is extraordinarily high at 86 per cent followed by agricultural engineering (42 per cent). Such a high intensity of migrants in these courses appear to be due to limited number of institutions offering course (say, in the case of aeronautical engineering) and competitive examination that forces students to out migrate.

Conclusions and Policy Imperatives

Agriculture has recorded abysmally low growth during last two years and this is a matter of serious concern for the sustainability of overall growth and food security. Agriculture has been confronted with numerous bottlenecks both from demand and supply sides and declining public investment has raised serious concern for the long-term growth. Education and skill are critical for this sector as substantive proportion of our workforce in agriculture are either illiterate or having schooling up to primary level. In the fast changing economic milieu and WTO regime, a more serious challenge is how to improve skills and technical education. High level manpower is a most critical factor for research and development and enhancing inventive capacity. However, expansion of institutions has been tardy and qualitatively much lower. Institutional capacity of intake is low and there has been gross mis-utilisation of human resource. There has been disparity of institutions across regions. Almost one third agricultural graduates are unemployed thereby raising concerns of their quality. Emoluments are much lower compared to other disciplines and it is no better than average salary of primary school teacher. If we have to launch a second green revolution then we need to expand our capacity of agricultural and allied manpower both in terms of numbers and also in terms of quality in order to become more agile and adaptable in responding to the fast changing external environment. There is, therefore, need for urgent revamping of agriculture education and research policy.

References

1. Bhalla, G.S. (2006) "Globalisation and Agricultural Policy in India", in *Indian Agriculture in the New Millennium: Changing Perceptions and Development Policy* (Eds: N.A.Mazumdar and Uma Kapila), Academic Foundation, Vol.2.
2. Biswas, Gautam (2010), "Faculty Related Issues", in *Profile of Engineering Education in India: Status, Concerns and Recommendations*, Narosa Publishing House Pvt Ltd, New Delhi.
3. Chadha, G.K. (2006), "Indian Agriculture in the New Millennium: Human Response to Technology Challenges" in *Indian Agriculture in the New Millennium: Changing Perceptions and Development Policy*, (Eds: N.A.Mazumdar and Uma Kapila),

Academic Foundation, Vol.1.

4. Chakravarty, Sukhmoy (1988), "The Logic of Employment-Oriented Planning in Developing Countries", G.L. Mehta Memorial Lecture delivered at Madras.

5. Government of India (2010), *Economic Survey*, 2009–10, Ministry of Finance.

6. Indian Council of Agricultural Research (2010), *Proceedings of the Conference of Vice-chancellors of Agricultural Universities and Interface with ICAR Directors*, 17–18 February.

7. Jadhav, Narendra (2010), "Education System has a Long Way to Go", *The Statesman*, June 11.

8. Jha, C.S. (2010), "Challenges Faced by Indian Engineering Education", in *Profile of Engineering Education in India: Status, Concerns and Recommendations*, Narosa Publishing House Pvt Ltd, New Delhi.

9. Papola, T.S. (1993), "Labour Market Monitoring for Employment Planning: A Framework Illustrated with the Indian Experience" in *Labour Market and Human Resource Planning in Asia: Perspectives and Evidence*, (Eds: M.Maqtada and A. Hildeman), ILO/ARTEP, New Delhi.

10. Rao, C.H. Hanumantha (2006), "Liberalization of Agriculture in India: Some Major Issues" in *Indian Agriculture in the New Millennium: Changing Perceptions and Development Policy*, (Eds: N.A.Mazumdar and Uma Kapila) Academic Foundation, Vol.1.

11. *Rao, U.R. (2003), Report on "Revitalizing Technical Education"*, Ministry of Human Resource Development.

12. Singh, Neena (2006), Restructuring LIS User Education Courses in Universities of Agricultural Sciences: A study, *Annals of Library and Information Studies*, Vol. 53, September 2006, pp. 134–1 42, September.

13. Vaidyanathan, A. (2010), *Agricultural Growth in India: Role of Technology, Incentives and Institutions*, Oxford University Press, New Delhi.

14. World Education News and Reviews, (WENR, 2007), *Engineering Education in India: A Story of Contrasts*, Vol.20, No.1, January, 2007.

Table: 1 : Number of Institutions Engineering Degree Courses over the years

States/UTs	1990		1995		2000		2005		2008	
	Degree	*Diploma*	*Degree*	*Diploma*	*Degree*	*Diploma*	*Degree*	*Diploma*	*Degree*	*Diploma*
NORTH										
Jammu & Kashmir	2	6	2	11	5	12	5	12	NA	NA
Himachal Pradesh	1	5	2	6	2	6	5	6	5	7
Uttar Pradesh	NA	NA	35	90	89	100	117	120	NA	NA
Uttarakhand	NA	NA	NA	NA	NA	NA	NA	NA	NA	NA
Punjab	7	16	10	31	13	31	53	36	58	36
Haryana	6	13	11	19	29	21	43	33	NA	NA
Total	**16**	**40**	**60**	**157**	**138**	**170**	**223**	**207**	**63**	**43**
%	**5.9**	**5.91**	**14.32**	**18.07**	**17.01**	**16.68**	**15.91**	**18.05**	**10.13**	**7.06**
SOUTH										
Kerala	9	29	16	42	33	56	87	55	NA	NA
Karnataka	52	163	55	174	87	185	125	180	NA	NA
Tamil Nadu	41	130	77	141	156	210	258	214	275	227
Andhra Pradesh	27	78	32	84	107	91	267	101	NA	NA
Total	**129**	**400**	**180**	**441**	**383**	**542**	**737**	**550**	**275**	**227**
%	**47.60**	**59.08**	**42.96**	**50.75**	**47.22**	**53.19**	**52.60**	**47.95**	**44.21**	**32.27**

States/UTs	1990		1995		2000		2005		2008	
	Degree	*Diploma*	*Degree*	*Diploma*	*Degree*	*Diploma*	*Degree*	*Diploma*	*Degree*	*Diploma*
NORTH EAST										
Assam	3	9	4	9	4	10	4	10	6	10
Manipur	NA	NA	NA	NA	NA	NA	1	1	NA	NA
Meghalaya	NA	NA	NA	NA	NA	NA	1	1	NA	NA
Sikkim	NA	NA	NA	NA	NA	NA	NA	NA	NA	NA
Mizoram	NA	NA	NA	NA	NA	NA	0	1	NA	NA
Tripura	NA	NA	NA	NA	NA	NA	1	1	1	1
Arunachal Pradesh	NA	NA	NA	NA	NA	NA	1	1	1	1
Nagaland	NA	NA	NA	NA	NA	NA	0	1	NA	NA
Total	**3**	**9**	**4**	**9**	**4**	**10**	**8**	**16**	**8**	**12**
%	**1.10**	**1.33**	**0.95**	**1.35**	**0.49**	**0.98**	**0.57**	**1.34**	**1.29**	**1.97**
EASTERN										
West Bengal	12	30	15	32	33	39	52	45	NA	NA
Orissa	6	16	9	24	28	29	42	32	NA	NA
Bihar	NA	NA	NA	NA	NA	NA	NA	NA	NA	NA
Jharkhand	NA	NA	NA	NA	NA	NA	NA	NA	NA	NA
Total	**18**	**46**	**24**	**56**	**61**	**68**	**94**	**77**	-	-
%	**6.64**	**6.79**	**5.73**	**6.44**	**7.52**	**6.67**	**6.71**	**6.71**		

States/UTs	1990		1995		2000		2005		2008	
	Degree	*Diploma*	*Degree*	*Diploma*	*Degree*	*Diploma*	*Degree*	*Diploma*	*Degree*	*Diploma*
WEST										
Goa	NA	NA	NA	NA	NA	NA	NA	NA	NA	NA
Gujarat	NA	NA	NA	NA	NA	NA	NA	NA	NA	NA
Rajasthan	NA	NA	NA	NA	NA	NA	NA	NA	NA	NA
Maharashtra	85	132	123	147	166	155	178	190	201	293
Total	**85**	**132**	**123**	**147**	**166**	**155**	**178**	**190**	**201**	**293**
%	**31.36**	**19.50**	**29.36**	**16.92**	**20.47**	**15.21**	**12.70**	**16.56**	**32.31**	**48.11**
CENTRAL										
Madhya Pradesh	10	31	17	37	33	45	71	46	NA	NA
Chhattisgarh	4	6	4	7	11	10	15	10	21	30
Total	**14**	**37**	**21**	**44**	**44**	**55**	**86**	**26**	**21**	**30**
%	**5.16**	**5.46**	**5.01**	**5.06**	**5.42**	**5.40**	**6.14**	**2.67**	**3.38**	**4.92**
UTS										
Delhi	6	13	7	15	15	19	21	19	NA	NA
Andaman Nicobar	NA	NA	NA	NA	NA	NA	NA	NA	NA	NA
Chandigarh	NA	NA	NA	NA	NA	NA	5	4	5	4
Dadar Nagar Haveli	NA	NA	NA	NA	NA	NA	NA	NA	NA	NA

States/UTs	1990		1995		2000		2005		2008	
	Degree	*Diploma*	*Degree*	*Diploma*	*Degree*	*Diploma*	*Degree*	*Diploma*	*Degree*	*Diploma*
Daman Diu	NA	NA	NA	NA	NA	NA	NA	NA	NA	NA
Lakshadweep	NA	NA	NA	NA	NA	NA	NA	NA	NA	NA
Puducherry	NA	NA	NA	NA	NA	NA	NA	NA	NA	NA
Total	**6**	**13**	**7**	**15**	**15**	**19**	**26**	**23**	**5**	**4**
%	**2.21**	**1.92**	**1.67**	**1.73**	**1.85**	**1.86**	**1.86**	**2.00**	**0.80**	**0.66**
Grand Total	**271**	**677**	**419**	**869**	**811**	**1019**	**1401**	**1147**	**622**	**609**

Source: *NTMIS Nodal Centre, IAMR (various issues)*

Table: 2 Intake of Agriculture and Allied Engineering Graduates by States

States/Union Territories	2001		2002		2003		2004		2005		2006		2007		2008	
	Actual	*Sanct.*	*Actual*	*Sanct.*	*Actual*	*Sanct.*	*Actual*	*Sanct.*	*Actual*	*Sanct.*	*Actual*	*Sanct.*	*Actual*	*Sanct.*	*Actual*	*Sanct.*
Andhra Pradesh	40	40	40	40	40	40	40	40	60	60	110	110	NA	NA	NA	NA
Arunachal Pradesh	17	30	18	30	15	30	12	30	12	30	15	30	8	30	NA	NA
Assam	0(837)	0(918)	0(982)	0(955)	0(901)	0(955)	0(913)	0(963)	0(928)	0(1002)	NA	NA	NA	NA	NA	NA
Bihar	NA	NA	NA	NA	NA	NA	NA	NA	NA	NA	NA	NA	NA	NA	NA	NA
Chhattisgarh	0(3138)	NA	0(3361)	NA	0(3401)	NA	0(3654)	NA	0(4073)	NA	0(4252)	NA	NA	NA	NA	NA
Goa	NA	NA	NA	NA	NA	NA	NA	NA	NA	NA	NA	NA	NA	NA	NA	NA
Gujarat	NA	NA	NA	NA	NA	NA	NA	NA	NA	NA	NA	NA	NA	NA	NA	NA
Haryana	62	55	54	115	119	175	89	195	80	195	NA	NA	NA	NA	NA	NA
Himachal Pradesh	0 (236)	0 (250)	0(570)	0(710)	0(873)	0(940)	0(1029)	0(1190)	0(1163)	0(1210)	0(1181)	0(1270)	0(1283)	0(1350)	0(1505)	0(1530)
Jammu & Kashmir	0(733)	NA	0(795)	NA	0(745)	NA	0(849)	NA	NA	NA	NA	NA	NA	NA	NA	NA
Jharkhand	NA	NA	NA	NA	NA	NA	NA	NA	NA	NA	NA	NA	NA	NA	NA	NA
Karnataka	225	232	170	232	232	232	188	225	NA	232	NA	232	NA	NA	NA	NA
Kerala	20	22	50	52	53	55	26	60	NA	NA	NA	NA	NA	NA	NA	NA
Madhya Pradesh	90	90	108	110	99	110	102	110	99	110	80	110	18	110	NA	NA
Maharashtra	692	757	719	757	745	757	NA	223	NA	223	NA	227	NA	227	NA	NA
Manipur	0 (66)	0 (115)	0(86)	0(115)	0(88)	0(115)	0(80)	0(115)	0(81)	0(115)	0(91)	0(115)	0(102)	0(115)	NA	NA
Meghalaya	0 (109)	0 (180)	0(139)	0(180)	0(147)	0(180)	0(154)	0(180)	0(156)	0(180)	0(157)	0(180)	0(150)	0(180)	NA	NA
Mizoram	NA	NA	NA	NA	NA	NA	NA	NA	NA	NA	NA	NA	NA	NA	NA	NA
Nagaland	NA	NA	NA	NA	NA	NA	NA	NA	NA	NA	NA	NA	NA	NA	NA	NA

States/Union Territories	2001		2002		2003		2004		2005		2006		2007		2008	
	Actual	*Sanct.*	*Actual*	*Sanct.*	*Actual*	*Sanct.*	*Actual*	*Sanct.*	*Actual*	*Sanct.*	*Actual*	*Sanct.*	*Actual*	*Sanct.*	*Actual*	*Sanct.*
Orissa	49	60	49	60	32	100	84	160	106	130	106	130	130	130	130	130
Punjab	109	115	113	115	116	115	100	115	120	145	130	135	111	151	NA	NA
Rajasthan	NA	NA	NA	NA	82	115	145	220	124	230	NA	NA	NA	NA	NA	NA
Sikkim	NA	NA	NA	NA	NA	NA	NA	NA	NA	NA	NA	NA	NA	NA	NA	NA
Tamil Nadu	NA	120	NA	120	NA	120	112	120	111	200	NA	NA	NA	NA	NA	NA
Tripura	0 (126)	0 (120)	0(120)	0(120)	0(139)	0(160)	0(167)	0(160)	0(161)	0(160)	0(250)	0(250)	0(253)	0(250)	NA	NA
Uttar Pradesh	NA	380	NA	380	NA	380	NA	550	NA	550	NA	NA	NA	NA	AN	NA
Uttarakhand	NA	NA	NA	NA	NA	NA	NA	NA	NA	NA	NA	NA	NA	NA	NA	NA
West Bengal	346	361	470	631	659	711	715	711	766	792	NA	NA	NA	NA	NA	NA
Delhi	65	30	30	30	30	30	78	90	35	60	NA	NA	NA	NA	NA	NA
Andaman Islands	NA	NA	NA	NA	NA	NA	NA	NA	NA	NA	NA	NA	NA	NA	NA	NA
Chandigarh	26	20	16	20	23	20	28	20	20	20	13	20	16	26	25	30
Dadar Nagar Haveli	NA	NA	NA	NA	NA	NA	NA	NA	NA	NA	NA	NA	NA	NA	NA	NA
Daman Die	NA	NA	NA	NA	NA	NA	NA	NA	NA	NA	NA	NA	NA	NA	NA	NA
Lakshadweep	NA	NA	NA	NA	NA	NA	NA	NA	NA	NA	NA	NA	NA	NA	NA	NA
Puducherry	NA	NA	NA	NA	NA	NA	NA	NA	NA	NA	NA	NA	NA	NA	NA	NA
Total	**1741**	**2312**	**1837**	**2692**	**2245**	**2990**	**1719**	**2869**	**1533**	**2977**	**454**	**994**	**510**	**447**	**155**	**160**

* *includes agriculture engineering, dairy technology, environmental engineering, food technology, forestry, fishery (including marine engineering), Jute and sugar technology*

Source: *NTMIS Nodal Centre, IAMR (various issues).*

Table: 3 Outturn of Agriculture and Allied Engineering and Total Graduates by State

States/Union Territories	2001		2002		2003		2004		2005		2006		2007		2008	
	A&A	Total	A&A	Total	A&A	Total	A&A	Total	A&A	Total	A&A	Total	A&A	Total	A&A	Total
Andhra Pradesh	39	10258	36	14646	31	20165	34	20238	49	26870	65	31841	NA	NA	NA	NA
Arunachal Pradesh	5	149	16	155	12	133	16	131	10	123	11	125	10	108	15	155
Assam	0	542	0	574	0	808	0	692	0	688	NA	NA	NA	NA	NA	NA
Bihar	NA	NA	NA	NA	NA	NA	NA	NA	NA	NA	NA	NA	NA	NA	NA	NA
Chhattisgarh	0	752	0	934	0	1336	0	1422	0	1964	0	2277	0	2004	NA	NA
Goa	NA	NA	NA	NA	NA	NA	NA	NA	NA	NA	NA	NA	NA	NA	NA	NA
Gujarat	NA	NA	NA	NA	NA	NA	NA	NA	NA	NA	NA	NA	NA	NA	NA	NA
Haryana	46	1792	35	3209	34	4545	25	6112	23	6546	NA	NA	33	8157	NA	NA
Himachal Pradesh	0	251	0	299	0	298	0	260	0	215	0	426	0	331	0	560
Jammu & Kashmir	0	158	0	118	0	393	0	558	NA	NA	NA	NA	0	385	NA	NA
Jharkhand	NA	NA	NA	NA	NA	NA	NA	NA	NA	NA	NA	NA	NA	NA	NA	NA
Karnataka	134	14173	110	14195	150	18600	124	24821	141	29615	127	27848	48	33601	NA	NA
Kerala	16	4089	15	3724	17	3944	20	6554	NA	NA	NA	NA	27	9377	NA	NA
Madhya Pradesh	45	4479	28	5595	60	6090	68	7243	58	8553	6	11217	18	10374	2	11306
Maharashtra	291	26341	317	26791	331	27157	271	38850	297	39292	300	40885	51	42537	NA	NA
Manipur	NA	NA	NA	NA	NA	NA	0	65	0	40	0	33	0	31	0	41
Meghalaya	NA	NA	NA	NA	NA	NA	0	23	0	35	0	36	0	14	0	16
Mizoram	NA	NA	NA	NA	NA	NA	NA	NA	NA	NA	NA	NA	NA	NA	NA	NA
Nagaland	NA	NA	NA	NA	NA	NA	NA	NA	NA	NA	NA	NA	NA	NA	NA	NA

States/Union Territories	2001		2002		2003		2004		2005		2006		2007		2008	
	A&A	Total	A&A	Total	A&A	Total	A&A	Total	A&A	Total	A&A	Total	A&A	Total	A&A	Total
Orissa	30	2991	30	3453	31	5067	32	5152	50	6555	50	7029	52	7717	82	8542
Punjab	89	1981	91	2812	91	2938	85	3474	92	4599	92	6024	98	7439	98	7066
Rajasthan	NA	NA	NA	NA	33	1976	67	4496	77	6169	129	7494	NA	NA	NA	NA
Sikkim	NA	NA	NA	NA	NA	NA	NA	NA	NA	NA	NA	NA	NA	NA	NA	NA
Tamil Nadu	40	16670	79	20550	75	28107	80	41348	76	47916	NA	NA	NA	NA	NA	NA
Tripura	0	112	0	108	0	138	0	158	0	144	0	151	0	167	0	154
Uttar Pradesh	307	4862	367	6723	328	8118	365	14759	364	20417	NA	NA	180	19267	NA	NA
Uttarakhand	NA	NA	NA	NA	NA	NA	NA	NA	NA	NA	NA	NA	NA	NA	NA	NA
West Bengal	238	2856	216	3528	211	4873	223	7040	356	8748	NA	NA	21	12161	NA	NA
Delhi	0	974	27	1382	20	2126	26	2743	27	2557	NA	NA	0	3361	NA	NA
Andaman Nicobar	NA	NA	NA	NA	NA	NA	NA	NA	NA	NA	NA	NA	NA	NA	NA	NA
Chandigarh	NA	NA	NA	NA	16	409	11	446	5	423	12	838	13	831	20	925
Dadar Nagar Haveli	NA	NA	NA	NA	NA	NA	NA	NA	NA	NA	NA	NA	NA	NA	NA	NA
Daman Die	NA	NA	NA	NA	NA	NA	NA	NA	NA	NA	NA	NA	NA	NA	NA	NA
Lakshadweep	NA	NA	NA	NA	NA	NA	NA	NA	NA	NA	NA	NA	NA	NA	NA	NA
Puducherry	NA	NA	NA	NA	NA	NA	NA	NA	NA	NA	NA	NA	NA	NA	NA	NA
Total	**1280**	**92730**	**1367**	**108796**	**1440**	**137221**	**1447**	**186585**	**1625**	**211469**	**792**	**136224**	**431**	**156901**	**217**	**28765**

Note: Agriculture and allied (A&A) includes agriculture and allied engineering graduates.

Source: *NTMIS Nodal Centre, IAMR (various issues).*

Table 4 : Number of Teachers in Agricultural Engineering Degree Courses, 2005-06

Sl. No	States/UTs	Actual	Sanctioned
1	Andhra Pradesh	12 (9966)	17 (13922)
2	Arunachal Pradesh	6 (43)	8 (74)
3	Assam	0 (224)	0 (316)
4	Bihar	NA	NA
5	Chhattisgarh	0(635)	0(858)
6	Goa	NA	NA
7	Gujarat	NA	NA
8	Haryana	38 (1839)	NA
9	Himachal Pradesh	0 (361)	0(427)
10	Jammu & Kashmir	0 (258)	NA
11	Jharkhand	NA	NA
12	Karnataka	42 (9526)	42(11400)
13	Kerala	26 (4322)	26 (4638)
14	Madhya Pradesh	16 (2487)	NA
15	Maharashtra	74 (10993)	NA
16	Manipur	0 (45)	0 (73)
17	Meghalaya	0 (39)	0 (42)
18	Mizoram	NA	NA
19	Nagaland	NA	NA
20	Orissa	41 (1633)	49 (1999)
21	Punjab	25 (3292)	31 (3358)
22	Rajasthan	33 (1854)	39 (2035)
23	Sikkim	NA	NA
24	Tamil Nadu	39 (13089)	45 (13917)
25	Tripura	0 (58)	0 (58)
26	Uttar Pradesh	36 (1239)	NA
27	Uttarakhand	NA	NA
28	West Bengal	203(2860)	216 (3177)
29	Delhi	0 (1076)	NA
30	Andaman Nicobar	NA	NA
31	Chandigarh	29 (257)	32 (325)
32	Dadar Nagar Haveli	NA	NA
33	Daman Diu	NA	NA
34	Lakshadweep	NA	NA
35	Puducherry	NA	NA

Note: *Figures within parenthesis are total teachers.*

Source: *NTMIS Nodal Centre, IAMR.*

Table 5: State-wise Activity Statuses of Engineering Degree Holders, 2007 Batch

State	Paid Employed	Self Employed	Apprentices	Unemployed	Studying	Other Activities	Total
Chandigarh	366	5	0	64	78	0	513
%	71.35%	0.97%	0.00%	12.48%	15.20%	0.00%	100
Delhi	2552	0	8	668	133	0	3361
%	75.93%	0.00%	0.24%	19.88%	3.96%	0.00%	100
Haryana	5569	69	0	1961	558	0	8157
%	68.27%	0.85%	0.00%	24.04%	6.84%	0.00%	100
Himachal Pradesh	227	2	0	77	25	0	331
%	68.58%	0.60%	0.00%	23.26%	7.55%	0.00%	100
Jammu & Kashmir	343	0	0	11	31	0	385
%	89.09%	0.00%	0.00%	2.86%	8.05%	0.00%	100
Punjab	1671	64	0	1568	935	0	4238
%	39.43%	1.51%	0.00%	37.00%	22.06%	0.00%	100
Rajasthan	6614	102	0	1257	1281	187	9441
%	70.06%	1.08%	0.00%	13.31%	13.57%	1.98%	100
Uttar Pradesh	13724	0	104	4968	471	0	19267
%	71.23%	0.00%	0.54%	25.79%	2.44%	0.00%	100

State	Paid Employed	Self Employed	Apprentices	Unemployed	Studying	Other Activities	Total
Andhra Pradesh	NA	NA	NA	NA	NA	NA	NA
%	-	-	-	-	-	-	100
Karnataka	12541	62	332	16217	3472	977	33601
%	37.32%	0.18%	0.99%	48.26%	10.33%	2.91%	100
Kerala	5052	6	136	2887	1296	0	9377
%	53.88%	0.06%	1.45%	30.79%	13.82%	0.00%	100
Tamil Nadu	NA	NA	NA	NA	NA	NA	NA
%	-	-	-	-	-	-	100
Arunachal Pradesh	96	1	1	4	3	3	108
%	88.89%	0.93%	0.93%	3.70%	2.78%	2.78%	100
Assam	739	0	0	33	15	3	790
%	93.54%	0.00%	0.00%	4.18%	1.90%	0.38%	100
Manipur	12	0	2	14	3	0	31
%	38.71%	0.00%	6.45%	45.16%	9.68%	0.00%	100
Meghalaya	9	0	0	2	1	2	14
%	64.29%	0.00%	0.00%	14.29%	7.14%	14.29%	100
Mizoram	0	0	0	0	0	0	0
%	-	-	-	-	-	-	100
Nagaland	0	0	0	0	0	0	0
%	-	-	-	-	-	-	100

State	Paid Employed	Self Employed	Apprentices	Unemployed	Studying	Other Activities	Total
Orissa	0	0	0	0	0	0	0
%	-	-	-	-	-	-	100
Tripura	92	4	5	56	5	5	167
%	55.09%	2.40%	2.99%	33.53%	2.99%	2.99%	100
West Bengal	10435	151	138	643	698	96	12161
%	85.81%	1.24%	1.13%	5.29%	5.74%	0.79%	100
Gujarat	0	0	0	0	0	0	0
%	-	-	-	-	-	-	100
Madhya Pradesh	9280	0	0	846	292	0	10418
%	89.08%	0.00%	0.00%	8.12%	2.80%	0.00%	100
Chhatisgarh	1082	0	2	593	318	9	2004
%	53.99%	0.00%	0.10%	29.59%	15.87%	0.45%	100
Maharashtra	25346	93	93	13732	3190	83	42537
%	59.59%	0.22%	0.22%	32.28%	7.50%	0.20%	100
All India	**95750**	**559**	**821**	**45601**	**12805**	**1365**	**156901**
%	61.03%	0.36%	0.52%	29.06%	8.16%	0.87%	100

Source: *NTMIS Nodal Centre, IAMR.*

Table 6 : Discipline-wise and Gender-wise Distributions of Unemployed Engineering Degree Holders of 2007 Batch in India

DISCIPLINE	UNEMPLOYED ENGINEERING DEGREE HOLDERS											
	Looking for a Job			Not looking for a job			Interested in Self-employment			Total		
	Male	*Female*	*Total*	*Male*	*Female*	*Total*	*Male*	*Female*	*Total*	*Male*	*Female*	*Total*
Agricultural Engineering	103	39	142	0	1	1	0	0	0	103	40	143
All India	32349	12804	45153	147	120	267	157	24	181	32653	12948	45601

Source: *NTMIS Nodal Centre, IAMR.*

Table 7 : Discipline and Gender-wise Distribution of Paid Employed Engineering Degree Holders according to sectors, 2007

Discipline	SECTORS OF EMPLOYMENT														
	Central Govt.		*State Govt.*		*Local Body*		*PSU*		*Private Sector*		*Others*		*Total*		
	Male	*Female*	*Male*	*Female*	*Male*	*Female*	*Male*	*Female*	*Male*	*Female*	*Male*	*Female*	*Male*	*Female*	*Total*
Agricultural Engineering	3	1	10	13	2	0	27	1	107	30	30	3	179	48	227
Total	1784	598	2239	956	2839	960	8838	2454	52986	18209	3022	865	71708	24042	95750

Source: NTMIS Nodal Centre, IAMR.

Table 8 : Distribution of Employed Engineering Degree Holders of 2007 Batch in India by Discipline and Sector of Employment

Discipline	SECTORS OF EMPLOYMENT														
	Central Govt.		*State Govt.*		*Local Body*		*PSU*		*Private Sector*		*Others*		*Total*		
	Male	*Female*	*Male*	*Female*	*Male*	*Female*	*Male*	*Female*	*Male*	*Female*	*Male*	*Female*	*Male*	*Female*	*Total*
Aeronautical engineering	0	0	0	0	0	0	6	0	16	4	0	0	22	4	26
Agricultural engineering	3	1	10	13	2	0	27	1	107	30	30	3	179	48	227
Architectural engineering	4	6	9	10	53	23	49	58	686	750	68	42	869	889	1758
Biotechnology	5	11	28	7	19	2	56	26	352	240	24	8	484	294	778
Chemical engineering	53	13	62	23	28	6	253	39	888	240	82	37	1366	358	1724
Civil engineering	106	33	190	62	213	34	352	55	2660	594	158	27	3679	805	4484
Computer engineering	513	272	632	411	825	518	2532	1578	17833	7670	1125	454	23460	10903	34363
Electrical engineering	250	60	371	118	296	79	1163	152	4171	1442	147	41	6398	1892	8290

Discipline	SECTORS OF EMPLOYMENT														
	Central Govt.		*State Govt.*		*Local Body*		*PSU*		*Private Sector*		*Others*		*Total*		
	Male	Female	Male	Female	Male	Female	Male	Female	Male	Female	Male	Female	Male	Female	Total
Electronics and communication engineering	549	164	493	252	647	255	2434	459	14967	5325	597	124	19687	6579	26266
Mechanical engineering	198	10	289	22	671	19	1367	51	8788	1275	376	5	11689	1382	13071
Metallurgy engineering	9	3	2	0	2	0	21	2	127	21	5	0	166	26	192
Mining engineering	0	0	17	0	0	0	31	0	80	16	3	0	131	16	147
Industrial/ production engineering	22	2	28	12	44	3	70	8	589	105	148	15	901	145	1046
Textile engineering	1	0	30	0	6	0	37	4	220	53	23	10	317	67	384
Others	71	23	78	26	33	21	440	21	1502	444	236	99	2360	634	2994
Total	1784	598	2239	956	2839	960	8838	2454	52986	18209	3022	865	71708	24042	95750

Source: NTMIS Nodal Centre, IAMR.

Table 9: Discipline-wise Average Monthly Emoluments of Engineering Degree Holders of 2007 Batch in India

DISCIPLINE	EMOLUMENTS (Rs.)					
	Fresher			Experienced		
	Maximum	*Minimum*	*Average*	*Maximum*	*Minimum*	*Average*
Aeronautical engineering	38000	24000	30000	38000	24000	30000
Agricultural engineering	21033	8662	13949	22900	11000	16510
Architectural engineering	23083	11258	15000	24000	12900	15299
Biotechnology	25357	10742	17783	31000	18000	24554
Chemical engineering	30480	14473	21347	31000	18000	23651
Civil engineering	29300	13000	21000	30000	17000	21800
Computer engineering	35000	10000	18269	35465	13500	24824
Electrical engineering	29531	13256	20121	30000	15875	20166
Electronics and communication engineering	36855	10537	20494	37566	10970	21308
Mechanical engineering	33918	12641	20980	34250	13583	21247
Metallurgy engineering	33927	17061	23770	34750	26250	25767
Mining engineering	22808	12000	16768	25500	21500	17612
Industrial/ production engineering	28833	12033	17032	31500	12500	18520

DISCIPLINE	EMOLUMENTS (Rs.)					
	Fresher			Experienced		
	Maximum	*Minimum*	*Average*	*Maximum*	*Minimum*	*Average*
Textile engineering	19093	9322	25849	25833	11700	27962
Other streams	29563	14834	20829	30125	16466	25598
All India	**31173**	**11324**	**16094**	**31558**	**11854**	**20767**

Source: *NTMIS Nodal Centre, IAMR.*

Table 10 : Employed Degree Holders of 2007 Batch in India by Place of Work and Discipline

Sl.	DISCIPLINE	PLACE OF WORK						TOTAL	
		Within the State		Outside the State		Abroad			
		Male	*Female*	*Male*	*Female*	*Male*	*Female*	*Male*	*Female*
1	Aeronautical engineering	0	0	19	4	3	0	22	4
2	Agricultural engineering	95	40	83	8	1	0	179	48
3	Architectural engineering	662	736	203	144	4	9	869	889
4	Biotechnology	229	157	250	137	5	0	484	294
5	Chemical engineering	815	204	538	151	13	3	1366	358
6	Civil engineering	2378	551	1195	242	106	12	3679	805
7	Computer engineering	11706	5915	11568	4953	186	35	23460	10903
8	Electrical engineering	2811	842	3543	1030	44	20	6398	1892
9	Electronics and communication engineering	10061	3898	9457	2654	169	27	19687	6579
10	Mechanical engineering	6138	697	5360	679	191	6	11689	1382
11	Metallurgy engineering	59	7	97	19	10	0	166	26
12	Mining engineering	42	0	89	16	0	0	131	16

Sl.	DISCIPLINE	PLACE OF WORK						TOTAL	
		Within the State		Outside the State		Abroad			
		Male	*Female*	*Male*	*Female*	*Male*	*Female*	*Male*	*Female*
13	Industrial / production engineering	691	118	204	27	6	0	901	145
14	Textile engineering	208	52	108	15	1	0	317	67
15	Others	568	417	1721	213	71	4	2360	634
All India		**36463**	**13634**	**34435**	**10292**	**810**	**116**	**71708**	**24042**

Source: *NTMIS Nodal Centre, IAMR.*

Table 11 : State wise Distribution of Migrant Engineering Degree Holders of 2007 Batch who have Migrated for Education

State	*Total Out turn*	*Migrants other states*	*Residents of the state but obtained degree from elsewhere*	*Net Migration (Col.3-4)*	*Net Outturn (Col.2+5)*
1	2	3	4	5	6
Chandigarh	513	511 (99.61)	2(0.39)	509	1022
Delhi	3361	286(8.51)	497(14.79)	-211	3150
Haryana	8157	1306(16.01)	0	1306	9463
Himachal Pradesh	331	124(37.46)	30(9.06)	94	425
Jammu & Kashmir	385	0	12(3.12)	-12	373
Punjab	4238	596(14.06)	26(0.61)	570	4808
Rajasthan	9441	737(7.81)	0()	737	10178
Uttar Pradesh	19267	386(2.00)	88(0.46)	298	19565
Andhra Pradesh	NA	NA	NA	NA	NA
Karnataka	33601	2594(7.72)	12(0.04)	2582	36183
Kerala	9377	338(3.60)	312(3.33)	26	9403
Tamil Nadu	NA	NA	NA	NA	NA
Arunachal Pradesh	108	86(79.63)	0()	86	194

State	*Total Out turn*	*Migrants other states*	*Residents of the state but obtained degree from elsewhere*	*Net Migration (Col.3-4)*	*Net Outturn (Col.2+5)*
1	**2**	**3**	**4**	**5**	**6**
Assam	790	297(37.59)	0()	297	1087
Manipur	31	2(6.45)	0()	2	33
Meghalaya	14	11(78.57)	0()	11	25
Mizoram	NA	NA	NA	NA	NA
Nagaland	NA	NA	NA	NA	NA
Orissa	NA	NA	NA	NA	NA
Tripura	167	10(5.99)	0()	10	177
West Bengal	12161	731(5.48)	375(3.08)	356	12517
Gujarat	NA	NA	NA	NA	NA
Madhya Pradesh	10418	667(6.40)	98(0.94)	569	10987
Chhatisgarh	2004	46(2.29)	5(0.25)	41	2045
Maharashtra	42537	1554(3.67)	0()	1554	44091
All India Total	**156901**	**10282(6.55)**	**1457(0.93)**	**8825**	**165726**

Source: *NTMIS Nodal Centre, IAMR*

Table 12 : Discipline wise Distribution of Migrant Engineering Degree Holders of 2007 Batch who have Migrated for Education

Courses	*Total Out turn*	*Migrants other states*	*Residents of the state but obtained degree from elsewhere*	*Net Migration (Col.3-4)*	*Net Outturn (Col.2+5)*
1	2	3	4	5	6
Aeronautical engineering	29	21(72.41)	0	21	50
Agricultural engineering	431	48(11.14)	8 (1.86)	40	471
Architectural engineering	2185	184 (8.42)	31 (1.42)	153	2338
Biotechnology	1742	177 (10.16)	7 (0.40)	170	1912
Chemical engineering	3003	279 (9.29)	50 (1.67)	229	3232
Civil engineering	7411	527 (7.11)	90 (7.11)	437	7848
Computer engineering	53777	2714 (5.05)	356 (0.66)	2358	56135
Electrical engineering	13973	1137(8.14	108 (0.77)	1029	15002
Electronics and Communication engineering	42684	2551 (5.98)	322 (0.75)	2229	44913
Mechanical engineering	22899	1685 (7.36)	239 (1.04)	1446	24345
Metallurgy engineering	296	85(28.72)	19 (6.42)	66	362
Mining engineering	199	38 (19.10)	3 (1.51)	35	234

Courses	*Total Out turn*	*Migrants other states*	*Residents of the state but obtained degree from elsewhere*	*Net Migration (Col.3-4)*	*Net Outturn (Col.2+5)*
1	2	3	4	5	6
Industrial / production engineering	2146	170 (7.92)	14 (0.65)	156	2302
Textile engineering	761	56 (7.36)	4 (0.52)	52	813
Others	5365	610 (11.37)	206 (3.84)	404	5769
Total	**156901**	**10282(6.55)**	**1457(0.93)**	**8825**	**165726**

Source: *NTMIS Nodal Centre, IAMR*

8

Employability of Engineering Graduates

*Dr.M.P. Chandrasekharan

ABSTRACT

This paper examines the issue of employability of fresh engineers. Employability depends on knowledge, skills and abilities, all these three can be attained through education, vocational training and life skill training. Acquisition of knowledge has no short cuts. Today's teaching institutions (and Universities) give more emphasis on teaching and ignore the learning process. Familiarity and memory are tested in examinations. Continuous evaluation is disliked by students and teachers. Unless there is a desire to learn the whole process of studying is meaningless. What you learn becomes relevant on the job if learning is application oriented and digested by the learner. What kind of knowledge does the employer look for? DOMAIN KNOWLEDGE: In the modern jargon the employer wants the job seeker to have knowledge of his specialised field. Domain knowledge is the compatibility between the label and the content (A civil engineer should know civil engineering, a chemist chemistry etc). Skills are generally of two

*Dr.M.P. Chandrasekharan, Dean(Engineering), Amrita Viswa Vidyapeedam, Coimbatore,TN.

types:(a) Skill for identification and solution of problems Technical skills to perform routine work efficiently. The first one is associated with acquisition of knowledge and the second on the level of training. Part of the Ability is intrinsic, and part acquired. Some Intrinsic abilities are learning ability, Initiative, Confidence, Team Spirit, Organizational skill and Attitude. Acquired abilities include, Linguistic abilities, Communication skills, Organizational ability, Presentation skills (Spoken and written), Time Management, Stress Management, Personality grooming, Ability for group discussions, Critical Thinking and Etiquette sate.

Introduction : The Context

In India, nobody can be blamed if one sets one's sole objective of education as employment. In spite of lofty ideals set by educationists and institutions on academic excellence, the fact remains that university students, by and large crave a lucrative employment on securing a degree. One of the reasons why education in Arts and Science suffered a set back, and pure learning is unheard of, is this race among students for employment, motivated by their parents and society. Engineering colleges have mushroomed near every bus stop, and the student population had an explosive growth in recent times. Under the circumstances, it is only natural that a good percentage of engineering students are incapable of going through the rigorous academic program of professional education. There is always a pressure on institution to reduce the academic standards to suit the capability of the incoming students. It is only natural that this inward spiral would ultimately result in producing large numbers of graduates with hollow or non-content degree.

On the positive side, industrial growth of the country has been impressive, particularly in the IT sector, and employment is plenty. Today, for most graduates, getting a job is not a problem, while getting the right job is. Of late, employers have become more discerning and very choosy. Gone are the days when a graduate earns a job just by his label. In spite of the unemployment in the country, we find a large number of graduates who are unemployable. It is in this context that this paper on Employability is presented here.

Employability

Employability is a person's capability of gaining initial

employment, maintaining employment and obtaining new employment.

A graduate cannot be employed just because he has a degree in a specific area. There are three essential attributes that the person should have -

- Knowledge
- Skills
- Abilities

Traditionally, Universities and Colleges concentrate on imparting knowledge, and imagine that the student would acquire the skills himself, and that the abilities are inborn qualities. In the modern context, the Institution has to be proactive in imparting all the three requirements to make a student employable. However, the techniques adopted for imparting knowledge, skills and abilities ought not to be the same. In general it is thought that education brings about knowledge, vocational training, the skills and life skill training, the abilities.

Knowledge : Acquisition Strategies

Acquisition of knowledge has at least five different levels as follows :

- Familiarity with the topic
- Memory and reproduction
- Analysis
- Synthesis
- Creativity

Of the 40 theory courses and 16 practical courses done in B.Tech. Curriculum, one does not expect the student to reach the level of creativity in all of them. But the tragedy of today is that, learning of all subjects is reduced to the level of memorizing and reproducing it in the examination. Even in courses where the student should reach the level of synthesis and creativity (like design subjects and project) certain shortcuts are employed to deflect the student from going anywhere above the level of memory and reproduction. In fact this is the tragedy of our educational system in the acquisition of knowledge.

An employer expects the student to have domain knowledge.

An electronics engineer should know electronics. There should be a correlation between the label and content.

There is a perception among many employed people that the job requirement has nothing to do with what they learnt in college, and therefore, college imparts useless knowledge. This is true in the case of many people who enter the profession and do not grow in their technical domain. The rules of thumb they learn on the job attain predominance over fundamental knowledge and they think that these rules of thumb are enough, and should replace fundamental knowledge and education. Such an approach would justify McAwley who wanted to create a generation of workers and "babus" while the real thinking is done back home in England. In a modern nation like India, it is foolish to continue that policy. However, it is also important to appreciate that bookish knowledge alone does not make a person employable. Theoretical knowledge, complemented by practical skills should form the "Domain Knowledge" of any professional specialization. There are many who think that Theoretical and Practical are opposites; they are not. The opposite of "Practical" is not theoretical, but "Impractical".

Acquisition of Skills

Skills can be classified into two categories namely (a) skill for identification and solution of problems (b) Technical skills to perform routine work efficiently. The first type of skill is of a higher intellectual order whereas the second is close to the labour skill. Therefore, the first type of skill is closely associated with acquisition of knowledge. The second type depends on the level of training.

Employers generally complain that Universities and Colleges emphasize more on knowledge and less on skills. The universities in turn argue that they cannot reduce themselves to training schools for a specific industry. Universities should produce graduates who can adapt themselves to a variety of situations and cannot lose generality in imparting of knowledge and skills. The industry, on the other hand should provide adequate training to adapt the graduates to their specific situations, realizing that university education is not a substitute for on the job training.

On the part of the Universities, the whole concept of "Practical Classes" should change. Today it has deteriorated to a level where the student gains almost nothing from such classes.

Abilities

Today, institutions spend a considerable amount of money and effort on preparing the students to face interviews, thinking that this barrier is the most difficult one to get across. Abilities have different components, many are intrinsic and quite a few acquired.

Intrinsic Abilities

- Learning Ability : This is ingrained with the personality of the student, and ought to be sharpened at an early age
- Initiative : Although a student can mimics this quality by imitating his peers and teachers this is a basic quality of the individual
- Confidence : Level of confidence can be improved by facing different and difficult problems, and solving them oneself
- Passion : This quality is purely intrinsic
- Sense of ownership: When a certain work is entrusted would the student do it as if it were his own.
- Team spirit: Basically this quality is intrinsic. Very selfish people will have much less team spirit compared with others
- Organizational Skill: This is partly intrinsic, but the quality also can be enhanced by training, but basic skill should be present
- Attitude : This is purely intrinsic
- Values and Ethics : These qualities are acquired from the mother's lap

Acquired Abilities

The abilities listed below can, by and large be acquired by learning and through training. Most universities now emphasize on "Soft Skill Training" by imparting these abilities as a requirement of employability

- Linguistic abilities
- Communication skills
- Organizational ability
- Presentation skills (spoken and written)
- Time Management
- Stress Management

- Personality grooming
- Ability for group discussion
- Critical thinking
- Etiquettes

Summary

In summary, on the relative importance of Knowledge, Skill and Ability (K.S.A) for employment can be stated as follows:

- Knowledge without skill and ability is useless
- Skill without knowledge and ability is stale
- Ability without knowledge and skill is hollow

❑❑❑

9

Failure Rates and their Determinants in Engineering Degree Courses in Kerala

Factor Analysis of a Socially Deprived Group*

*K. Pushpangadan,

ABSTRACT

The paper examines the nature and causes of failure rates among students in engineering degree courses in Kerala for the period, 1991-2011. Analysis shows that there is a declining trend in pass rates whether it is government, semi-government, aided or unaided college. The distribution of pass rates among the social classes shows it is highly skewed; the students from SC&ST have the lowest pass rates and from the forward class the highest. The findings suggest that the skilled human capital in this segment of the labour market is highly inequitable in Kerala providing an explanation for the widening consumption inequality observed in the state relative to southern states in India. The determinants

*K. Pushpangadan,, Centre for Development Studies, Kerala

are multidimensional in nature belonging to social, economic, demographic and educational environment at home and college. We have undertaken a case study of the determinants of the failure rates among the most deprived class of student (ST and SC) who have completed the course work but not received the degree from MG university, during the period 1991- 2011.

Among several variables identified and measured, 14 variables were selected for final analysis due to the sample size. Principal component analysis (PCA) from Factor Analysis is used here as a data reduction technique among the fourteen variables. Screen plot test reduces the fourteen variables to seven factors (composite variables). The statistical significance of the factor shares suggest the following variables contributing much of the variation in the seven factors: (1) the percentage marks scored in mathematics, physics and English language in SSLC and Higher Secondary courses; (2) Teachers' support, (3) Family support; (4) Learning practices; and (5) Learning atmosphere. For policy purposes, these focal variables require further probing to single out the specific variables within each category. This is the future area of research.

Introduction

The first comprehensive study on the development policy of the regional economy of Kerala for United Nations had recommended, among other things, promotion of skill-intensive but low capital intensive industrial activity to absorb the increasing number of unemployed youth particularly the educated in the economy (UN, 2000; p. 5). Such a strategy presupposes that economic growth is dominated by industrial activity. But evidence clearly indicates that growth is instead driven by the service sector and continues to be so for the recent spurt of economic activity arising from economic liberalization. By all indications, the source of growth may not change in the future due to path dependence of growth process. Given the nature of growth, the employment generation for the region should have a two pronged strategy: one for the skill intensive industrial base at the national/international economy and the other for replacing the acute labour shortages existing in the economy by upgrading technology and by institutional-social innovations. The task is by no means simple but complex and requires deliberate public intervention in the skill formation in the educated labour force in the state. A basic requirement is the promotion of technical education with problem-solving skills needed for the shortages emerging in the state and the emerging demand for the national economy. In this context, the opening up of technical education to the private sector for increasing the number of engineering graduates is in the right direction provided there is no trade-off between quantity and quality of such an expansion.

The assumption that quality remains the same in the technical education is not valid for several reasons. The major ones are: (1) Pass rates are coming down especially after the expansion of self-financing colleges; (2) Widespread shortages of qualified teachers; and (3) No monitoring of teaching quality especially in the coverage of syllabus, its assessment and methods of teaching as envisaged by AICTE norms. All the issues need in depth analysis but we take up the pass/failure rates for detailed investigation to begin with.

The outline of the paper is as follows. Section 1 following the introduction, an analysis of the pass/failure rates is undertaken at the level of state, university and social class. Section 2 contains an assessment of factors that affect the performance. Section 3 deals with the Principal Component Analysis. Final section contains concluding remarks and suggestions for future research.

I

Pass/Failure rates: A trend analysis

State level

Figure 1. Outturn-intake ratio of engineering degree courses, Kerala,1991-2008

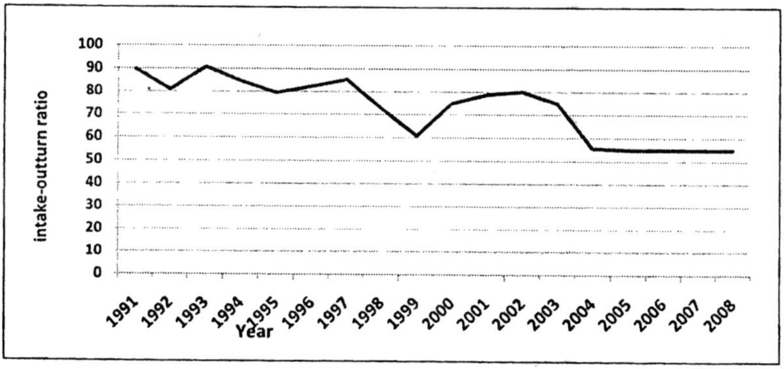

Source: Annual Technical Education Review (2003,2008)Note :1) Outturn-intake ratio = Outturn fig in(t+4)/Intake year in t

The trend analysis is undertaken for the period 1991-2008 using secondary source of data at state level. The outturn in any year is taken as the pass rate and it is normalized by the four-year lagged (backward) intake since the program is of four year duration. Actual outturn is not available for the years 2005, 2006, 2007, 2008 and we have assumed the same rate in 2004 for these years. The intakes for the years-1992, 1993 and 1994- are not available. So we have taken sanctioned capacity of students as its proxy and the pass rate, the outturn-intake ratio, is given in Figure 1. Figure 1 clearly indicates a declining trend. It may be noted that the initial pass rate was 90 % in 1991 which declined to 55% in 2008. The declining rate occurs at a time when the Government has sanctioned new colleges especially self-financing ones in large numbers. Let us examine the performance at the university level.

University level

Among the five major universities in the state, only Kerala University has provided the results in their website for the period 2006-2010. This information is examined below. Here the pass rate is taken as the ratio of the number of students passed in the final examination to the total number of students appeared for the final examination. In other words, the pass rate is not adjusted for the intake figures in any particular year. The details of the data base are given in Table 1.

The website covers a total of 19 colleges affiliated to Kerala University as shown in Table 1 which includes: two colleges under Government, one aided college, two Semi-Government college, one college under the direct management of the University and a sample of 15 of 35 self-financing colleges. The choice of self-financing colleges is limited to the colleges that have students appearing in the final examinations of the course group included in the website. The final sample size for the analysis included only nineteen colleges due to data availability. The pass rate in course group consisting of six popular branches (EC, EE, CE, IT, ME) by ownership pattern is given in Figure 2.

Table 1 : Sample of colleges by management and by branches, 2006-2011

Branch	No of colleges					Year/Periods coverage
	Govt.	*Semi Govt.*	*Aided*	*Self financing*	*Under direct Mgmt.*	
CE	1	0	1	5	0	2010(7)
CS	1	2	1	10	1	2006-10(15)
EC	2	2	1	10	1	2006-10(16)
EE	1	0	1	9	0	2009-10(11)
IT	1	1	0	7	1	2006-2008, 2010(10)
ME	2	1	1	8	0	2009-10(12)

Source: *www.keralauniversity.edu (accessed in 16 December, 2011)*

Note: *: CE- Civil Engineering; CS- Computer Science; EC - Electronics & Communication Engineering; EE- Electrical and Electronics Engineering; IT - Information Technology; and ME- Mechanical Engineering.*

Figure 2 : Pass rates in course group by type ownership affiliated to Kerala University, 2006-2010

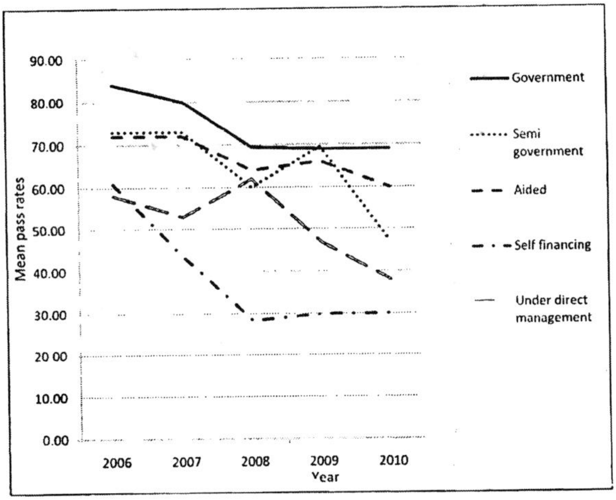

Source: *www.keralauniversity.edu (accessed in 16 December 2011)*

Figure 2 shows a declining trend in pass rates in the course group during the period, 2006-2010. The difference between the worst category (self-financing) and the best category (government) is 24% in 2006 and 39% in 2010. This indicates a widening gap between the two groups during the period. In the case of the aided college, the rate begins from 72 % in 2006 and declines to 60% in 2010. The gap is only 12 % in this category. The fluctuations are the highest in the case of semi-government colleges. The reasons for such variation in the performance need careful and urgent attention. The case of Govt. colleges particularly the oldest one (College of Engineering, Trivandrum) is a puzzle since they have the best of everything: quality of teachers and students, and infrastructure. Let us examine the change in the pass rates during the period by ownership pattern as given in Figure 3.

Figure 3 : Percentage Change in pass rates in a course group by institutional category, Kerala University 2006-2010

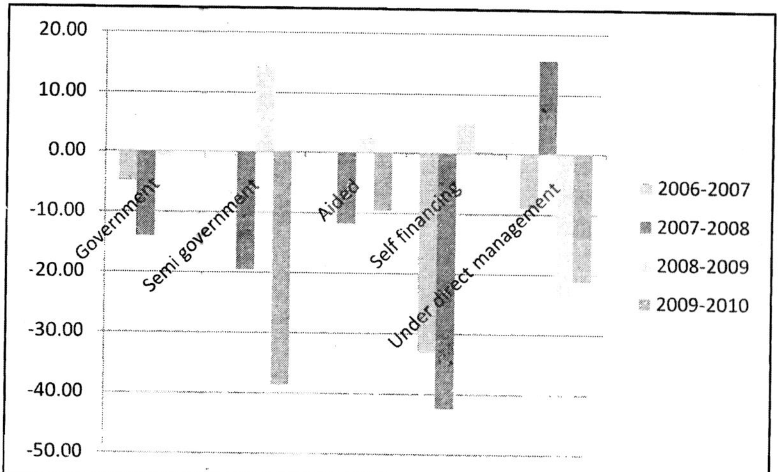

Source: www.keralauniversity.edu (accessed on 11December 2011)

In Figure 3 Govt. Colleges consistently show a negative change in pass rates unlike the other four categories. The positive changes occurred only once in the remaining four categories but not in the same year. For example, positive growth in the pass rate is observed in 2008-9 for three categories (semi-government, aided and self-financing) and in 2007-2008 for the fifth category (direct management).

The highest negative change is recorded in self-financing colleges followed by semi-Govt. colleges for the year 2007-2008. The reason for inter category variation in the performance needs careful study.

The observations on other universities in the state are as follows. It could be seen that sharp difference exists in pass percentage between colleges in the same university. For example, Model Engineering College, Thrikkakara had a pass percentage of 67.5 % in 2009 which could be contrasted with 16.6 % of Sarabhai Institute of Science and Technology, Thiruvananthapuram, both are affiliated to Cochin University of Science and Technology (CUSAT). The pass percentages remained low in most of the colleges under CUSAT. The situation in MG University is similar, if not worse. No information is available in the case of Calicut University. As a result, we are not in a position to judge its performance.

Performance of Social Category

The analysis so far has not examined the performance of social classes reflecting the share of human capital formation in technical courses in the state. This is taken up below.

Figure 4 : Pass out-intake ratio by Social Category, Kerala, 2004 Batch

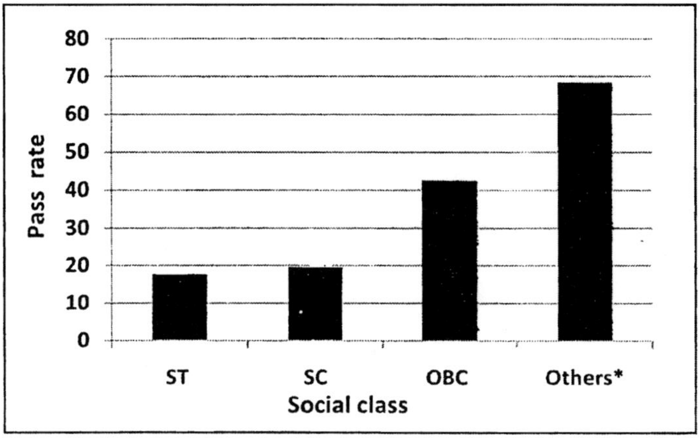

Source: Annual Technical Education Review 2006 & 2008

Notes : 1) Others include NRI;

2) Intake refers to the academic year 2004;

3) Pass-out refers to the year 2008.

The inequality in performance by social category is assessed only for the intake year 2004 at the state level due to data limitation. The lowest pass rate (17.7 %) is for ST and the highest (68.4 %) for others (non-reserved category). The pass rate of OBC is more than twice that of ST and SC. The rate of non-reserved class is approximately 3.5 times that of the socially deprived class (SC & ST). The analysis has the limitation of not matching the intake and the pass out for a given year. Let us examine the inequity after adjusting for this limitation.

Pass rate of most-preferred branches in selected colleges

The pass rate examined so far is not related to the intake in a particular year and its pass rates in the final examinations after four years. Such information is not available in published sources. To overcome this difficulty, we have traced the pass rates of students in 2010 from the intake of year 2006 in two most-sought after branches (Electronics and Communication, Electrical and Electronics) in five colleges from the primary records of MG University. The resulting pass rate, intake-outturn ratio, by social category is given in Figure 5.

Fig 5 : Intake-outturn ratio in two branches in five colleges by social category, MG University, 2006 Batch

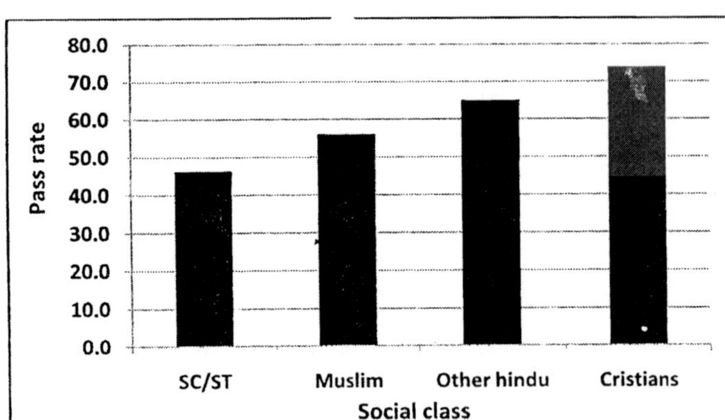

Source: Primary source, MG University

The state level analysis clearly shows that OBC performance is more than double that of SC/ST students but the proportion is not reflected in the two branches and five colleges for Muslim community where the improvement is marginal. This result has to be judged when community has better colleges with better infrastructure facility and teaching faculty. This again requires detailed investigation.

The distribution of skilled human capital as reflected in the pass rates of social category is highly skewed. The empirical evidence shows that share of socially and economically deprived classes, say Muslim community, with degree certificates is about 25 % points lower than the socially and economically better off communities in the state. The inequality in skill distribution will be translated into earnings inequality particularly in an economy where the lead sector of growth is tertiary sector. This provides a partial explanation of the widening consumption inequality in Kerala since liberalisation. A full-fledged explanation requires a detailed analysis of the other sources of consumption inequality in the region.

The empirical analysis indicates the need for undertaking studies on the determinants of the performance for public action. There are several casual observations made on the determinants in the literature on the phenomena. First, the determinants are social, economic, institutional educational. Second the issue needs multi-dimensional framework and appropriate statistical tools for indentifying the specific factors for public action to minimize the social wastage. This is taken up in the next section.

II

Determinants of Failure Rates

As mentioned earlier, there are a large number of variables that affect the performance of the students. After interaction with experts in the field, students and teachers, and social scientists, we have identified more than forty variables that may affect the performance. The questionnaire was finalized after pre - testing the same for the primary survey. The next step is to select the sample.

Selection of Sample

The sample selection is restricted to the most socially deprived class of students, SC & ST, in our society. We have asked MG University for helping us with the survey which they have readily agreed.

MG University had asked all SC and ST students who had completed the course work but not received the degree to report at the University campus for a primary survey by our study team. About eighty students from different schemes and branches of engineering reported for the survey. We have administered the questionnaire to them and collected the data. After reviewing the completed questionnaires for accuracy, we have selected only 72 of them for further analysis. But two of them belong to different social category and not included for the analysis for reasons of homogeniety. Hence the sample size is restricted to 70.

Measured variables

Of the forty variables measured in the primary survey, only 14 of them were selected for further analysis on the basis of quality of the data and to keep the general rule in factor analysis '...that the number of variables should be at least five times as many observations as there are variables to be analysed' (Hare et.al, 2005, p.99). The selected variables are: (1) one economic variable - occupation of the head of the household; (2) two variables on household characteristics - gender and total rooms in the household; (3) six measures on the performance of SSLC and Plus 2- marks awarded in mathematics in SSLC, marks awarded in mathematics in Plus 2, marks awarded in English language in SSLC, marks awarded in English language in Plus 2, marks in physics in SSLC and in Plus 2; and (4) five variables on the environment and infrastructure in which students are placed -- family support, teachers' support, place of stay, learning atmosphere and learning practices. These variables are measured either quantitatively or qualitatively.

The information from the primary survey is then subjected to factor analysis as discussed below.

III

The Empirical Results

Factor Analysis

The methodology used here is factor analysis, the multivariate statistical techniques of data reduction. Among the techniques available in factor analysis for data reduction, we opted for principal component analysis (PCA) which defines the basic dimensions of the data without making any assumptions about the common factors (Dhillon and Goldstein, 1984). More precisely, PCA '...is to determine the factors (i.e., principal components) in order to explain as much of the total variation in the data as possible with as few of these factors as possible' (Dillon and Goldstein, 1984; p.24). There are three steps in PCA: (1) Data adequacy test; (3) Extraction of the number of factors; and (3) Interpretation of the factors. The three stages of PCA are undertaken below.

Sample adequacy test

Two tests are given in the packages for judging the suitability of factor analysis in a given data set. They are: (1) Kaiser-Mayer-Olkin test and (2) Bartlett test of sphericity. It provides the statistical probability of significant correlation in among at least a few variable in the correlation matrix. The results are tabulated in Table 2.

Table 2 : Sample Adequacy test

1) Kaiser-Meyer-Olkin test		.566
2) Bartlett's Test of Sphericity	Approx. Chi-Square	180.094
	Df	91
	Sig.	.000

Source: Primary survey

Kaiser-Meyer-Olkin value is higher than .5, the data satisfy the sample adequacy test. The Bartleet's test also indicates statistical significance. Both sample adequacy tests are satisfied by the correlation matrix. Therefore we proceed to the next stage, the extraction of the number of factors for further analysis.

Extraction of the number of factors

The most commonly used technique for deciding the number of factors is the latent root criterion (LRC).The rationale for LRC is that any individual factor should account for the variance of at least a single variable if it is to be retained for interpretation. Each variable contributes a value of 1 to the total Eigen-value. In the literature, only the factors having latent root/Eigen value greater than or equal are considered significant. Graphically, Scree test criterion gives the optimum number of factors as given below.

Figure 6 : Eigen value plot for Scree test criterion

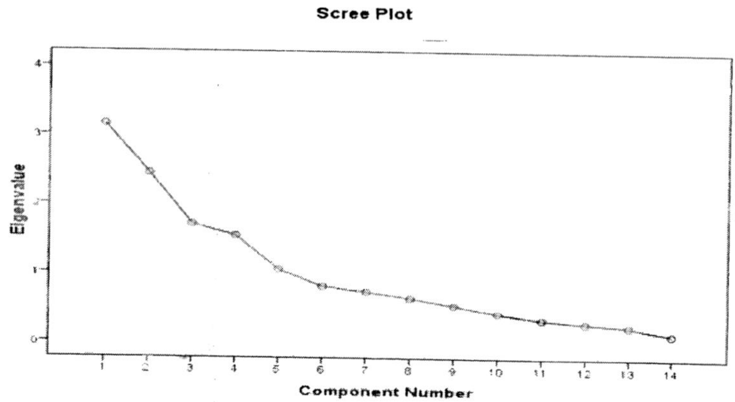

By this criterion, only five factors can be extracted (the component number corresponding to one is five). The literature suggests that if the variables are less than 50 then the factors extracted using LCR is conservative (Hare ed. al. 2005, p.104). To avoid such a conservative choice, the variance explained by the factors as in Table 3 can be used for deciding the final number of factors.

Table 3 indicates two more factors have Eigen value close to 1 which explains 82 % of the total variance. In other words, we extract seven factors for further analysis by LCR and total variance explained.

Interpretation of factor matrix

Interpreting the factor matrix depends on the final factor solution. The standard practice is to examine unrotated factor matrix based on orthogonal rotation for the seven factors. Acceptance of the factor matrix depends on the factor loadings (correlation of each variable and the factor) and their statistical significance. Let us examine the results of the unrotated factor matrix in Table 4.

The factor matrix should not have statistically significant cross-loading. Taking the significant level as more than 0.5, we have cross loadings in two variables, learning atmosphere and learning practices. In the case of the variable, learning atmosphere, the loadings are significant in factor 2 and 3. In the case of learning practices, statistically significant cross-loadings exist in factor 2 and

6. Therefore, the unrotated factor matrix cannot be used for interpretation of the structure of the variables that determine pass rates. We have to generate the factor matrix using rotated methods for interpretation. No rule has been developed to guide the researcher in selecting the method of rotation. There is no standard rotation method. Usually the researcher follows the method available in the statistical package. In our case, SPSS provides Varimax with Kaiser normalisation rotation method. We use the method and its factor matrix is reported in Table 5.

The factor matrix indicates that the following variables are statistically significant in the failure rates of the students surveyed. The important ones among them are ; (1) the quality of training in mathematics, physics and English at the secondary and higher secondary schools; (2) Family support; (3) Teacher's support;; (4) Place of stay; (5) Learning atmosphere and (6) Learning practices.

The study clearly indicates further research for identifying specific variables in each of the focal category for public action for reducing the gross wastage of financial resources and the time allocated for the course. This will be the next stage of our analysis.

Concluding Remarks

The pass rates show a declining trend in all subjects and in all colleges in the degree courses in engineering subjects during the period, 1991-2011. The pass rates of students from socially deprived classes are much lower than the students from socially advanced classes. The determinants of failure rates among most socially deprived classes of students are: (1) the quality of training in mathematics, physics and English at the secondary and higher secondary schools; (2) Family support; (3) Teacher's support; (4) Place of stay; (5) Learning atmosphere; and (6) Learning practices. These focal groups need detailed enquiry to isolate specific factors for public intervention.

The study also provides some interesting area for further research on 'how inequity in human capital formation contributes to the increase in the consumption inequality in Kerala observed in the recent past. There is an urgent need for detailed investigation that leads to specific factors enabling public action for reversing the low quality of technical education in the state.

Table 3. Total variance explained by principal components

Component	Initial Eigen values			Extraction Sums of Squared Loadings			Rotation Sums of Squared Loadings		
	Total	% of Variance	Cumulative %	Total	% of Variance	Cumulative %	Total	% of Variance	Cumulative %
1	3.15	22.5	22.5	3.15	22.5	22.5	2.11	15.04	15.04
2	2.44	17.44	39.94	2.44	17.44	39.94	2.02	14.41	29.45
3	1.71	12.19	52.13	1.71	12.19	52.13	1.83	13.08	42.53
4	1.53	10.94	63.06	1.53	10.94	63.06	1.65	11.76	54.29
5	1.04	7.46	70.52	1.04	7.46	70.52	1.39	9.91	64.19
6	0.81	5.77	76.29	0.81	5.77	76.29	1.35	9.64	73.83
7	0.73	5.22	81.51	0.73	5.22	81.51	1.08	7.68	81.51
8	0.64	4.58	86.09						
9	0.53	3.79	89.89						
10	0.42	3.02	92.9						
11	0.34	2.4	95.31						
12	0.28	2.03	97.34						
13	0.24	1.72	99.06						
14	0.13	0.94	100						

Source: *Primary survey*

Table 4 : Component matrix of principal component analysis (unrotated)

Component Matrix[a]

	COMPONENT						
	1	2	3	4	5	6	7
Occupation of Head of the Household	.247	-.306	.534	.507	-.152	.038	-.152
Gender	.144	-.656	-.113	.330	-.021	.552	-.080
Total Rooms	.192	.464	-.619	.055	-.266	-.038	.205
Total Marks for English (SSLC)	.760	.279	-.186	.270	.143	.170	.022
Total Marks for English (Plus2)	.828	.004	.173	.190	.098	.119	.069
Total Marks for Maths (SSLC)	.697	.287	-.406	-.022	-.169	.155	.005
Total Marks for Maths (Plus 2)	.399	-.048	.398	-.681	-.016	.146	.202
Total Marks for Physics (SSLC)	.485	.102	-.265	-.062	.587	-.235	-.348
Total Marks for Physics (Plus 2)	.711	-.009	.268	-.452	-.038	-.049	-.008
Family Support - Engineering	-.281	.551	.135	.381	.448	.032	.067
Teachers Support - Engineering	-.368	.706	.068	-.003	-.021	.387	.204
Place of Stay	.387	.287	.288	.450	-.431	-.405	.085
Learning Atmosphere	.034	.572	.622	.022	.179	.098	.019
Learning Practices	.131	-.563	-.055	.184	.355	-.175	.662

Extraction Method: Principal Component Analysis.

a. 7 components extracted.

Table 5 : Factor matrix of principal component analysis (VARIMAX rotation)

Rotated Component Matrix[a]

	COMPONENT						
	1	2	3	4	5	6	7
Occupation of Head of the Household	-.261	.038	-.006	.682	.002	.457	.020
Gender	-.008	-.095	-.291	-.035	-.051	.875	.146
Total Rooms	.791	-.191	-.030	-.072	-.077	-.275	-.046
Total Marks for English (SSLC)	.664	.168	.175	.253	.426	.250	.073
Total Marks for English (Plus2)	.367	.432	.079	.430	.353	.334	.178
Total Marks for Mathematics (SSLC)	.806	.234	-.094	.077	.216	.081	-.113
Total Marks for Mathematics (Plus 2)	-.049	.909	.004	-.101	-.063	-.046	.032
Total Marks for Physics (SSLC)	.176	.070	-.027	-.084	.890	-.070	.016
Total Marks for Physics (Plus 2)	.154	.805	-.134	.182	.245	-.038	-.036
Family Support - Engineering	-.041	-.367	.759	.015	.155	-.117	.053
Teachers Support – Engineering	.197	-.101	.719	-.230	-.374	-.137	-.230
Place of Stay	.253	.000	.056	.869	-.067	-.216	-.017
Learning Atmosphere	-.134	.277	.746	.248	.026	-.122	-.173
Learning Practices	-.081	.011	-.168	.008	.036	.130	.956

Extraction Method: Principal Component Analysis. Rotation Method: VARIMAX with Kaiser Normalization.

a. Rotation converged in 19 iterations.

References

* Prepared for the National Seminar on "Employability of Fresh Engineering Graduates in the Era of Globalisation: Problems and Prospects" at Cochin University of Science and Technology on 24-25 January, 2012. I place on record my appreciation to Dr. Sabu Thomas, Nodal Centre, Cochin University of science and technology, for inviting me to present the paper and chair one of the sessions in the conference. I am extremely thankful to P. Balakrishnan, Centre for Development Studies, P.K. Sivanandan, Institutte for Societal Advancement and Rajan Gurukkal, M.G. University for making this study possible. The paper in its present form owes a great deal to the discussions I had with P. Sivanadan, a co-investigator of the project on 'Failure rates and causes of engineering education in Kerala' undertaken by CDS on behalf of Government of Kerala. For excellent research and computer assistance, I am thankful to M. Rajesh and Vinitha Raj. The limitations, if any, of the paper are purely the author's.

1. Cochin University of science and Technology, Annual Technical Education Review, various issues (2003, 2006, 2008), Cochin.
2. Dillon, W.R.and M. Goldstein, 1984, Multivariate Analysis; Methods and Application (Wile, New York.
3. Hair, J.H., Anderson, R.A, Tatham, R.T and Black W.B, 2005, Multivariate Data Analysis, Fifth Edition (Pearson education, Delhi)
4. United Nations, 2000, Poverty, Unemployment and Development Policy: A case study of selected issues with reference to Kerala, Reprinted by the Centre for Development. Studies, Kerala, India.

10

Relevance of English Language Teaching in Technical Education Curriculum to face the Challenges of Globalisation

Ramesh U, Biju C V ** and Manoj R****

ABSTRACT

Globalization has influenced the educational philosophy and classroom practices all over the world. Over the past two decades, significant economic and political changes have affected the technical education system as such. The forces associated with globalization have altered the context in which educators operate, and profoundly changed one's experience in both formal and informal education. No doubt, technical education has attained a key position in the knowledge society under globalised economy. However, the challenges faced are immense and far-reaching. It is assumed that globalization promises dramatic and rewarding change to the technical education systems, of societies, which are

*Associate Professor & Placement Officer, College of Engineering Munnar.

**Associate Professor, College of Engineering Munnar.

*** Assistant Professor, College of Engineering Munnar.

relatively stable in their political, social and institutional makeup, while for the others, it may threaten the very stability needed to build well performing technical education system. Globalization and technological advancements are delivering an increasing access to the world and subsequently subjects should reflect this global outlook. The role of communication skills is now considered as the most crucial element in the make up of a Global Engineer. Communication levels at technical education not only require fluency in spoken language but also in written form. There exist ample evidences that graduate engineers lack the required standard of communication skills when compared to the needs posed by the industry in this globalised world. English language provides the students with ample opportunities to have a glimpse of the various developments in all spheres of human activity. Teaching of English language along with the technical curriculum, which ends up in the adaptation of English literature is a complex phenomenon as the learner and the target language belong to entirely different environment. The prevailing structure and methodologies is resulting in deteriorating the standards of particularly English Language Teaching because ELT has been always in clash with local culture of technical syllabi. Keeping the above in view, present paper deliberates on the impact of Globalization on Indian Technical Education System in general and also aims at revealing the need and role of educating English to technical students and how it could help in building up a successful career.

Technical Education in India

As on today we have more than 300 universities, institutions of higher learning and deemed universities, out of which 95 deemed to be universities, 13 institutions of national importance, 19 central universities, 203 state universities, 5 institutions established under state legislation act and about 16,885 colleges including 203 Autonomous colleges. Education System has increased fourteen-fold in terms of the number of universities and thirty three-fold in terms of the number of colleges, in comparison to the number at the time of Independence.

Technical education contributes a major share to the overall education system and plays a vital role in the social and economic development of our nation. In India, technical education is imparted at various levels such as: craftsmanship, diploma, degree, post-graduate and research in specialized fields, catering to various aspects of technological development and economic progress. The beginning of formal Technical Education in India can be dated back to the mid 19th Century. The major policy initiatives in the pre-independence period included appointment of the Indian Universities Commission in 1902, issue of the Indian Education policy resolution in 1904 and the Governor General's policy statement of 1913 stressing the importance of Technical Education, the establishment of IISc in Bangalore, Institute for Sugar, Textile and Leather Technology in Kanpur, N.C.E. in Bengal in 1905 and Industrial schools in several provinces.

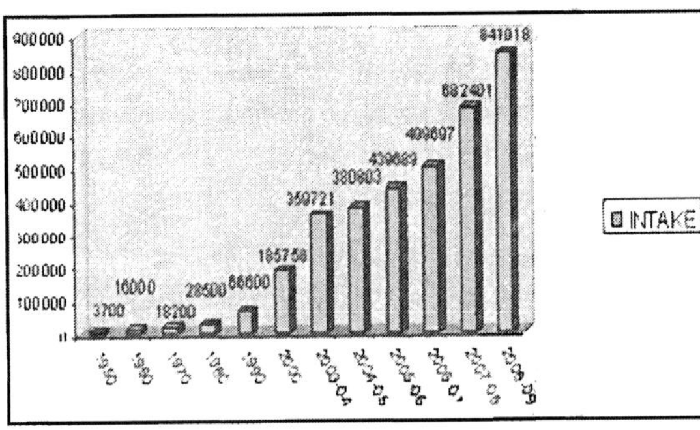

Sanctioned intake in Engineering Institutions

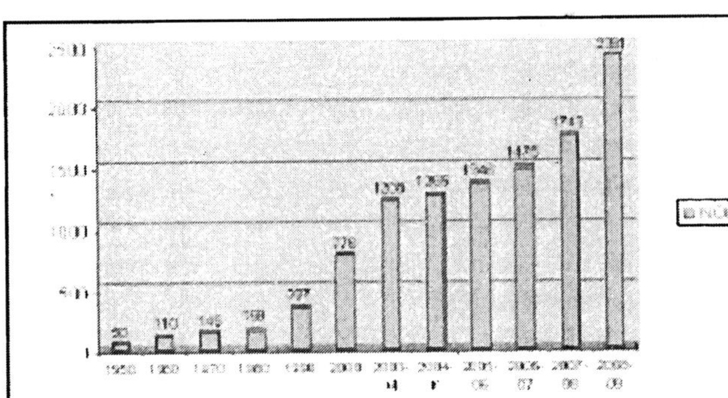

Growth of Engineering Institutions

All India Council for Technical Education (AICTE) was set-up in November 1945 as a national level Apex Advisory Body to conduct survey on the facilities on technical education and to promote development in the country in a coordinated and integrated manner. The growth of technical education before independence had been very small. The number of Engineering Colleges at the dawn of independence in 1947 was 44 with an intake of 2500. In the early eighties Govt. allowed private participation in the setting up of technical institutions on self financing basis. Following the same, a large number of privately managed institutions were established in the past 25 years and the growth is shown in the figure.

The compound annual growth rate of institutions during the 61 year period is 6.7% excluding the IIT's and NIT's and the percentage increase in the number of engineering institutions from 1995 to 2005 is 298.66%. The sanctioned intake has increased from 2940 in 1947 to 8.41 lakhs in 2008. The growth has brought along some serious concern of quality as well as efficiency.

Globalization and Education

As it is known, globalization, no doubt, promises a dramatic and rewarding change to the higher education systems of the developed countries. Where as for the developing and the underdeveloped countries, where the system is facing the scarcity of resource, it threatens the stability needed to build the well performing system. Developing countries often have to adjust willingly or unwillingly both to the quickening pulse of international change, and accordingly, reform on several fronts simultaneously, which may not be possible under the given resource status of higher education.

In spite of continued debates and discussions on this issue, still it is not clear whether this new global policy would keep in view the interests of global capitalism or the needs of particular societies in general and disadvantaged groups in particular. The fear that the process of Globalization Policy may apparently shape the perspectives of educational reforms in favour of those who have already benefited from the system as they are in the position to influence the policy is haunting those who are not in a position to influence the policy. Globalization is expected to be a process through which an increasingly free flow of ideas, people, goods, services and capital would lead to the integration of economies and societies. It is characterized by an accelerated flow of trade, capital, and information, as well as mobility of individuals, across geographical borders. It reflects a more comprehensive level of interaction than that has occurred in the past, suggesting something beyond the word "international". It implies a diminishing importance of national borders and strengthening of identities, that stretch beyond those rooted in a limited locale in terms of particular country or region. It can also be defined as the intensification of worldwide social relations which link distant localities in such a way that local happenings are shaped by events occurring at any distant place and vice versa. It is this construction of time-space compression that has given rise to popular notion of "One-World" "Global Village", etc.

Education is undergoing constant changes under the effects of globalization. The effects of Globalization on education bring rapid developments in technology and communications are foreseeing changes within school systems across the world as ideas, values and knowledge, changing the roles of students and teachers, and producing a shift in society from industrialization towards an information-based society. It reflects the effect on culture and brings about a new form of cultural imperialism. The rise of a global society, driven by technology and communication developments are shaping children, the future citizens of the world into 'global citizens', intelligent people with a broad range of skills and knowledge to apply to a competitive, information based society. The future of countries often lies within their ability to compete in a global market where industrial based economies are giving way to knowledge based industries, realising the importance of "knowledge, skills and the intellectual capacity to meet the challenges of accelerated change and uncertainty". Education is becoming a lifelong learning and training process, developing transferable skills and knowledge that can be applied to competitive markets where knowledge and information is being traded as a commodity. The notion of the world community being transformed into a global village, as introduced in 1960 by Marshall McLuhan in an influential book about the newly shared experience of mass media, was likely to be the first expression of the contemporary concept of globalization. Despite its entry into the common lexicon in the 1960s, globalization was not recognized as a significant concept until the 1980s, when the complexity and multidimensionality of the process began to be examined. Prior to the 1980s, accounts of globalization focused on a professed tendency of societies to converge in becoming modern, described initially by Clark Kerr and colleagues as the emergence of industrial man. One critical issue that emerges from all of these restructuring processes is the central role of knowledge, education and learning for the success of the Global Information Society (GIS) and global information economy. The Centre, State, Institutions and Regulatory Bodies have taken initiatives in order to face the challenges of globalization. The National Policy on Education 1986 and the revised policy, POA 1992 has made no reference to the operating of foreign universities in India or promotion of Indian Higher Education abroad. Infact there is a need for evolving a clear policy on these aspects of transnational education. However, in the recent past, an attempt has been made to work out the possibility of regulating the entry of the foreign universities. In doing so, UGC has recommended that foreign universities to offer their programmes in India; they should be accredited in their respective country; Degree awarded to the students should be valid for India as well as in the country where degree-giving institution is located; There should be reciprocal arrangements for Indian Universities to operate in those countries from where the institutions/programs have come to India.

Role of English Language

A critical issue to be debated under the phenomenon of globalization is the consequences for languages. To view globalization from a language perspective, we therefore need to consider the use and status of languages in a global context, or what Maurais calls "their relationships and their competition on the world's checkerboard". This approach might be called a sociolinguistic perspective on globalization. The organization of languages and their relationships in a global context are outlined by de Swaan (2001): The human species is divided into more than five thousand groups each of which speaks a different language and does not understand any of the others. With this multitude of languages, humankind has brought upon itself a great confusion of tongues. But nevertheless, the entire human species remains connected: the division is overcome by people who speak more than one language and thus ensure communication between different groups. It is multilingualism that has kept humanity, separated by so many languages, together. The multilingual connections between language groups do not occur haphazardly, but, on the contrary, they constitute a surprisingly strong and efficient network that ties together–directly or indirectly–the six billion inhabitants of earth. It is this ingenious pattern of connections between language groups that constitutes the global language system.

As a mode of communication, Latin transcended other communication venues as the formal language of the Roman Republic and Roman Empire. Later on, Latin was adopted by medieval scholars. For centuries, Latin served as the scholars' language. High schools and postsecondary schools around the world offered classes in Latin. Globally, this served as *the* common language between professors and students. Subsequently, German surpassed Latin and gained the language lead. *French speakers attempted to dominate the language world, as then the English.* With such diversification, Latin became relegated to philosophers and theologians. Schools, on a general scale, ceased teaching Latin. This historical leading language was limited to specialized schools. Journals published scientific articles in German, French and English. Because of an increased number of English-speaking universities in the United States, United Kingdom and other parts of the world, coupled with the United State's openness to international markets, the English language became essential. In order to communicate knowledge, English evolved as the common language not only for scholars, but for everyone participating in the international market.

The epicentre of the world's languages is English. English is the one language that connects the super central languages with one another and that therefore constitutes the pivot of the world language system. English obtained this position at the core of the global language system due to a variety of historical reasons, namely large scale migration and settlement of native language speakers, military imposition (colonialism), commercial or political power and prestige derived from scientific, cultural or other achievements. This language dynamic provided an advantage to American professors who published journals in their native tongue, under rules established by their peers, and with peer review from academicians from the same country. In a second instance, advantages were realized in world class universities located in some parts of the world, such as South Korea, India and Japan. To compete internationally, many countries decided to teach English as a second language. This was (and remains) the choice without exception. The English language subsequently opened doors of new opportunities for students to study in foreign countries. Some government actions in regards to mandated instruction in English provided many countries advantages in the long run.

English and India

In India, 100 million people know the English language. According to NationMaster.com, India is second in the world (only to the United States) in the number of English speakers by country. Hindi is the national language and primary tongue of 30% of India's people; yet there are 14 other official languages: Bengali, Telugu, Marathi, Tamil, Urdu, Gujarati, Malayalam, Kannada, Oriya, Punjabi, Assamese, Kashmiri, Sindhi, and Sanskrit. Hindustani is a popular variant of Hindi/Urdu spoken widely throughout northern India but is not an official language. English enjoys associate status in most areas, yet continues to be the most important language for national, political, and commercial communication. Evolution of the English language in India began when the British recruited local people to work in kitchens and in clerical jobs. Thus, the Indians learned English. In India, the independence struggle was first led by the 'babus' (English-educated Indians). A commonality of English language helped people network for their cause. English also supported establishment of a national spirit through a common mode of communication. Language trended toward native dialect usage under Mahatma Gandhi. For nearly 40 years, the emphasis was on education in mother tongues. In modern India, this has changed. The contemporary choice is, once again, English. This paradigm shift did not evolve as a result of government policies, but as a result of the perceptions of the Indian people. Especially for those who belong to high or medium classes, but also for working parents, Indians prefer that their children attend English schools.

In higher education, English continues to be the premier, prestige language. Careers in business and commerce, government positions of high rank (regardless of stated policy), along with science and technology (attracting many of the brightest) continue to require fluency in English. Fluency is also necessary for the many students who contemplate study overseas. The advent of computers encouraged English literacy as it became a common machine language. Jobs in information technology-oriented careers (computer software, hardware, and data entry) suddenly enhanced the necessity of knowledge of the English language. Working-class parents, themselves rural-urban migrants and perhaps bilingual in their village dialect and the regional standard language, perceive

English as the tool their children need in order to advance. The English speaker in India also encounters more courteous situational responses than does a speaker of an indigenous language.

Issues in English Language Teaching

English learner in India grows up wrapping a particular vernacular and endeavours to master a foreign language. Generally, his/her sources of English learning are Indian Teachers and Indian authored books. Till recently, regular academics in India comfortably identified communication skills with conventional terms like 'Business English', 'Technical English' and no stake holders complained. Such courses designed without feeling the pulse of the market naturally were mere academic exercises for filling the gap. The teachers of such programmes, mostly with the background of English literature, struggled to find some vestige of professional dignity both in the staff rooms and amongst the students.

Once the student enters the technical education system, the learner confronts writings of specialized area as one requires passing a specialized and prescribed syllabus. During this whole process, the learner has to undergo various stages of language learning which enable him to be proficient in the target language. It is a matter of pity that even students who have studied all along in English medium schools find it hard to communicate effectively in English. Another major problem faced by students who have spoken their native tongue for the major part of their life is to think in their native tongue. For instance, when a question is asked, they will first form the answers in the native language and then try to translate it into English. This will not positively work as it takes a lot of time and the translation process often gets messed up. Many Technical students find themselves better at written communication compared to oral communication. Intriguingly, a recent study found that poor English written communication skills tended to parallel poor written communication skills in the native tongue.

Keeping in view the role and importance of English, it is necessary that a teacher of English must be aware of the aims and objectives of teaching English. The basic aims of teaching English at the school level are to enable the students to understand English when spoken, to speak, to read and to write. Listening and Reading are passive or receptive skills whereas Speaking and Writing are active or productive skills. These aims can be classified in other terms also ie 'Reception' and 'Expression'. Reception involves understanding spoken and written English whereas Expression involves speaking and writing English. Unfortunately, the prevailing situation is resulting in deteriorating the standards of particularly English Language Teaching and a number of causes have been identified that are contributing directly or indirectly to the aggravation of the situation. The main causes that have stroked during the research are:-

- Vagueness of objectives of English teaching
- Lack of trained and efficient teachers
- Stress on theory rather than practicals
- Inappropriate approach of ELT
- Language should be acquired naturally.
- Inappropriate teacher student ratio.
- Alarming increase in the mushrooming of many substandard educational institutions in the recent past.
- ELT is done more as a subject than as a Language.

Overall, the teacher has a significant role to play. He should give the students as many opportunities as possible to listen and speak English. The teacher should become a good model of speech for the students. In order to become a good model, the teacher should himself learn and practice English sounds. While teaching English for practical purpose, students must be given the ability to write simple and correct English. This ability calls for handwriting, spellings, structures, punctuations and so on.

Conclusion

The Industrial and commercial life of the country is now to a major extend ruled by English Language. There have been decade's long efforts to reduce the geographical disparity bringing every youth to a common communication platform. Written and verbal communication skills are of paramount importance for any professional student. Once a student gets admitted to a professional institution, their books are in English, the media of instruction is English, all the examinations are in English and the seminars & presentations are to be carried out in English. It goes without saying that no matter how intelligent and knowledgeable a person is, no matter how many degrees he may possess, but he shall not be successful in getting into an appropriate career until and unless he possesses excellent and effective communication and listening skills. What should be included and excluded in the technical curriculum puts a lot of responsibility on the curriculum setters as it has been observed that in the past few decades , nothing significant has been done other than mere cosmetic changes at the graduate and post graduate levels. So keeping in view the serious issues regarding the standardized language teaching, confronted by the country, it becomes necessary for the Government authorities, education managers and students to realize their social responsibility in the direction to improve upon the standards of English Language Teaching.

11

Self Financing Colleges and Employability of Engineering Graduates

*Dr. Usha Menon

ABSTRACT

The paper highlights some survey reports. One report discussed the needs to focus on the development of skills of the graduates and suggested that the college should define the set of skills that a graduate is supposed to have after each semester. The colleges have to change pedagogical style from teacher-centric to student-centric, and include more assignments for students in order to analyse independently and to apply required tools on real life problems. It has also discussed that the faculty should define the set of skills that a graduate is supposed to have after every semester. This paper seeks to highlight the methods for implementing possible solutions without going through the impediments that colleges usually face.

Much has been spoken about enhancing the employability skills of engineering graduates in our state. A major challenge facing the

*Dr. Usha Menon, Jyothi Engineering College, Cheruthuruthy,Thrissur

engineering faculty is to produce engineering graduates who have more than just the traditional technical ability and skills. We need to add business acumen and managerial know-how to our students, for only then can they contribute positively to the industrial and business development. Employers are now seeking graduate engineers who have good communication skills and team working ability. However it is difficult to develop these skills adequately in an already packed curriculum. A recent Royal Academy of Engineering report identified six key attributes as being of particular importance to engineering employers. (J. King, "Educating Engineers for the 21st.Centuary"TheRoyal Academy of Engineering U. K., and June 2007) these are :

- Practical application of theory to real industrial problems
- Theoretical understanding
- Creativity and understanding
- Team working
- Technical breadth
- Business skills

According to a survey released by Indo Asian News Service which was jointly carried out by the Federation of Indian Chambers of Commerce and Industry (FICCI) and the World Bank 64 percent of surveyed employers are "somewhat","not very" or "not at all" satisfied with the quality of engineering graduates. According to Andreas Bloom,senior education economist at the World Bank "the survey highlights the need for colleges to focus on the skills of the graduates"

The state of Tamil Nadu, our closest neighbor evolved some strategies after considering the pros and cons of the matter. The Department of Higher Education came out with the following recommendations which have been considered seriously. I place before you the matter which was taken up by the Government of the state. I quote their official document before you.

- Source Coding Technical Education – Establishment of "Finishing School" in Government, Government Aided and in some Self Financing Engineering Colleges – To enhance the employability of students – Orders – Issued.

Higher Education (J2) Department

G.O.(Ms) No. 144 Dated: 6.5.2008. Chithirai 24,

Thiruvalluvar Aandu 2039.

Read:

From the Commissioner of Technical Education, Chennai-25 letter No.5294/Y3/SO(CDC)/2008, dated 12.2.2008.

ORDER:

In the letter read above, the Commissioner of Technical Education has sent the proposal for the establishment of "Finishing School" in Government, Government Aided and Self Financing Engineering Colleges in order to enhance the employability of Engineering Students. He has stated that the objectives of "Finishing School" programme are as follows :

(1) to enhance the employability of engineering students.

(2) to enhance the standard of fresh engineering graduates so that they become acceptable to the industry.

(3) to cater to the needs of the industry.

(4) to create a platform where the students can work towards accessing the skills required to get into industry.

(5) to bridge the gap between industry and educational institutions.

(6) to provide both technical and soft skills to people to facilitate their employability.

(7) to meet the needs of unemployed and non-employable engineering graduates and to improve their communication as well as I.T. skills.

(8) to reinforce the students skills and acquire industry-specific knowledge from trained faculty and experts from industries.

(9) to offer suitable candidates to different industries.

(10) to help young graduates to find jobs through appropriate training.

He has also stated that the 'Finishing School' programmes are learning-oriented than teaching-oriented and the recent boom in the Indian Economy has enhanced the need for such Finishing Schools.

2. The Government have examined the above proposal of Commissioner of Technical Education and decided to accept it. They accordingly direct that the Finishing School be started in Government, Government Aided and in some selected Self Financing Engineering

Colleges for all branches of Engineering on self supporting basis from May 2008 as detailed below:-

Programmes.

Students are to choose one course from the list of technical courses and one course from the list of soft skills mentioned below:-

I.T. related technical courses: (50 hours)

- Telecommunication Fundamentals
- Data Communication
- Telecommunication Networks and Switching
- Wireless and Cellular Communication
- Internet Services
- Network Management Systems
- Data base Management
- Management Information System

Soft Skills – Courses: (30 hours)

- Personality Development
- English Fluency and Communication Skills
- Counselling
- Public Speaking

Lab – (20 hours)

Target Group

(i) Final year outgoing students of Engineering Colleges

(ii) Unemployed pass out students from Engineering Colleges

Center

Finishing School courses will be conducted in all Government, Government Aided and some selected Self Financing Engineering Colleges. Maximum number of students admitted / College: 100

Curriculum

The curriculum and syllabi for Finishing Schools will be designed and prepared on par with IIT and NIT. Finishing School syllabi by the experts from the academia with intensive participation of experts from the industry and the stress is on making the students more suitable to the industry needs. A well planned curriculum will consist of the right mix of courses in Telecommunication, Computer, Networking, Software, Management and Soft Skills.

Course Fee

Course fee is Rs.1200/- for general candidates

50% concession (Rs.600/-) on course fee for Scheduled Castes/ Scheduled Tribes and physically challenged candidates.

Duration

Total Duration – 5 weeks

- 4 hours per day, 20 hours/week, 100 hours/ 5 weeks
- Technical Course = 50 hours
- Soft Skills = 30 hours
- Lab = 20 hours

Total =100 hours

Time Schedule

The programme will be run for a period of 5 weeks during May to June, 5 days/week, 4 hours/day with a total input for 100 hours.

Selection for Admission

First come first served basis.

Eligibility

Students from all branches of Engineering who have completed the degree (or) Final year out going degree students from all branches of engineering.

Evaluation Criteria

For each subject feedback and test shall be administered to evaluate the effectiveness of learning. Based on the students performance in test, certificate will be issued by the Directorate of Technical Education, Chennai-25.

(By order of the Governor)

K.GANESAN,

SECRETARY TO GOVERNMENT.

To

The Commissioner of Technical Education, Chennai-25.

The Vice-Chancellor, Anna University, Chennai, Coimbatore, Tiruchirappall, Tirunelveli.

Principal of all Government/Government Aided Engineering Colleges (through the Commissioner of Technical Education, Chennai-25)

Copy to

The Senior P.A. to Hon'ble Minister (Higher Education), Chennai-9.

SF/SC.

// forwarded / by order //

Section

In our own state the High court had the following remarks to make concerning the engineering teaching situation in the state,specially when dealing with self financing colleges. I quote from an article which appeared in the Times of India regarding the matter:

When quality takes a beating

The Kerala High Court has highlighted the issue of poor quality of education in engineering colleges in the State.

The reality: With many engineering colleges lacking qualified teachers, students have to fend for themselves.

Lack of qualified teachers in engineering colleges across Kerala seems to have opened up a serious debate on the need for stepping up the quality of education offered to students in these colleges. The issue assumes significance especially in the wake of the recent decision of the United Democratic Front government to give approval for starting 19 new engineering colleges. The college managements will have to fill many faculty posts in various engineering streams.

The court reached the decision on the basis of a report by an expert committee that found that the engineering colleges lacked qualified teachers. Investigations done by the committee revealed that only five of the 24 colleges inspected by it had qualified teachers.

Stipulation:

The Bench directed the State government to "critically scrutinize" the eligibility of every institution which had submitted applications seeking no-objection certificates. The court said that if the newly started colleges lacked any facilities, affecting the quality of education, the government should ensure that the students were allowed to join other colleges, given credit for the course study already undergone by them, and refunded the entire fee.

Explaining that the lack of qualified professionals continues to remain a major hurdle in the engineering education sector in the State, K.P.P. Pillai, former Executive Secretary of the Indian Society for Technical Education (ISTE), points out that there is no mandatory programme for improving the professional qualifications of engineering faculty members.

"Several teachers lack domain knowledge. Unless they spend a lot of time in the library and browse the Internet for new materials, their knowledge will not get updated. Most of the teachers also lack presentation skills. He says the fall in teaching standards is reflected in the decreasing pass percentage of engineering students in colleges in the State.

Students also complain that some of the faculty members use Malayalam to convey ideas. Teachers say that there is no other way, as several students find it difficult to understand the lectures in English. This is also partially correct which can be resolved.

A serious shortage of M.Tech. seats in the State remains the major problem for not having qualified faculty in engineering colleges, says G.P.C. Nayar, president of the Federation of Associations of Managements of Unaided Professional Educational Institutions in India.

Poor pay:

Engineering college teachers say most self-financing colleges do not offer proper service and pay packages to them. Many colleges are now appointing teachers without adequate qualification for low salaries. Mr. Sebastian says this will destroy the quality of professional education in the State and adversely affect the future of students passing out from these institutions.

K. Vasudevan, Professor of Electronics and Dean, Faculty of Technology at the Cochin University of Science and Technology, says talented teachers, especially at the middle level, are very few in most colleges. The teachers will be both old and retired persons who may not have high exposure to latest topics and students who have just passed out from their college and joined as teachers

The issue of unemployability of engineering graduates came to the fore when Union Human Resource Development Minister Kapil Sibal flagged it on January 3 in his keynote address to the Indian Science Congress held at SRM University, near Chennai. "...We have over 3,000 engineering institutions and colleges across the country, which produce nearly five lakh engineering graduates. While this expansion has made engineering education accessible to a large number of people, the quality of education imparted is a matter of concern," he said. "I am informed," Sibal added, "that out of every 100 campus candidates interviewed, most reputed companies are able to recruit only 10 to 20. The strike rate varies industrywise. This, when there is no dearth of availability of jobs. Undoubtedly, we need better quality education, as also new courses, new content and new delivery standards."

While several Vice-Chancellors and those who run self-financing engineering colleges do not agree with Sibal's assessment that most of the engineering graduates are not employable, discussions with placement officers of IT, manufacturing and service sector companies do reveal that a large percentage of the engineering graduates are indeed not employable. The supply outstrips the demand. Every student who applies for admission to the 517 engineering colleges for 2011-12 will get a seat. These 517 colleges come under Anna University, Chennai, government engineering colleges and self-financing engineering colleges affiliated to Anna University of Technology in Chennai, Madurai, Tiruchi and Coimbatore. What is worrying educationists is that 32,000 engineering seats will fall vacant this year although a mere pass is the eligibility criterion for students belonging to the Most Backward Classes, the Scheduled Castes and the Scheduled Tribes. The cut-off is 45 per cent marks for students belonging to the Backward Classes and 50 per cent for open competition. Besides, there is the problem of employability of engineering graduates to be tackled.

Professor Mannar Jawahar, Vice-Chancellor of Anna University, took a balanced view of reports that only about 25 per cent of the engineering graduates are employable. "This is based on a

NASSCOM report. This situation obtained five years ago. It is not like that now," he asserted. He divided the engineering graduates into three categories: those who get good jobs soon after graduation, the underemployed, and the unemployed. "The underemployed, who may be working in some company or the other, may graduate to a good job after three or four years. The unemployed engineers will get a job after a couple of years," he said. He estimated that those who got good jobs soon after their graduation would be about 25 per cent.

S. Vaidhyasubramaniam, Dean, Planning and Development, SASTRA University, Thanjavur, said the assessment that only 25 per cent of the engineering graduates were employable was based on a study conducted by the McKinsey Global Institute in 2005. "It is too old to be repeated now because a lot of effort has been made in the past six years to improve the quality of engineering graduates," he said.

In an article in an English-language newspaper published from Chennai on July 27, Vaidhyasubramaniam said:

"The NASSCOM-McKinsey report, 2005, titled 'Extending India's Leadership of the Global IT and BPO Industries', has been extensively used to show in a poor light the quality of Indian engineering graduates. The genesis of this report is the McKinsey Global Institute study, which states that 'only 25 per cent of engineering graduates in India have the skills to be employed in IT jobs without prior training'. The extent to which this finding has been used is of critical concern."

The report was based on interviews with human resource managers of 10 multinational companies out of a total of 83, he said. "It is not fair to assess the overall quality of a country's engineering graduates' output based on responses from a diminutive HR community. Ten out of 83 is not a good sample," he argued. "Even if one were to assume that the finding was true in the context of the overall standard of Indian engineering education, to quote a 2005 finding repeatedly is tantamount to accepting that no effort has been made to improve the quality in the past six years," he said.

"A lot of collaborative effort has been taken by universities, engineering colleges and IT companies to enhance employability," he claimed. It is time NASSCOM commissioned a study with regard to Indian engineering education and the quality of the engineering graduates by including only the circuit branches such as the CSE, IT, ECE, EEE and EIE as the relevant population, he said. "This will show a better result than the one obtained by including all engineering graduates as though engineering graduates are made for IT industry alone," he argued. The Vice-Chancellor of a Anna University of Technology pooh-poohed reports that only about 20 per cent of the engineering graduates are employable. "The remaining graduates may not get jobs immediately. But they will definitely be employed after a few years," he said.

Each University approaches the development of these soft skills and commercial awareness in its own way. Some use career development departments to offer optional training in employability skills, while others use the work placement approach. There are also many who rely on a loose industrial partnership to encourage the student's professional development. A major constraint that colleges face is the heavy yet necessary syllabus that has to be handled in a limited time. Lack of personnel to handle the extra workload is yet another factor. While on the one hand the redrafting of the syllabus to include another type of input to enhance student skills is something to be thought about on the other hand we are aware of the huge amount of red tapism and other hurdles that have to be overcome to affect any such move. Here is where self-financing colleges can have an edge over their counterparts. We at Jyothi Engineering College are working at including PAT hours (Placement and Training hours) into our regular schedule or timetable. The idea is to incorporate the training for the acquisition of all the necessary skills that are required of the engineering graduates into the curriculum .It then becomes mandatory for all the students. A huge amount of groundwork was necessary before the implementation of the idea. I am showcasing my own place of work only to illustrate why and how it is possible for the self-financing engineering colleges to solve the problem of employability. This self-financing college has evolved a syllabus for each of the four years that the students spend in the campus for the completion of the engineering course.

The management had put aside a large amount of money to be utilized solely for skill development programmes. ***In the first year, which is tightly packed academically, does not have a separate Pat hour***. However paper presentations made in the classroom are used as a means to teach the necessity of positive body language, clarity of speech and acceptance of critical remarks. Selected teachers are given an orientation in this kind of skill teaching, since we understand that not everyone has the time or the inclination to follow this methodology. We have found positive response from the students and a lot of enthusiasm in their participation in such teaching.

The second year includes an intense and more focused soft skill training programme. The goal here is to develop communicative skills so that the student is able to:

a) Put forth an idea clearly and confidently.
b) Is made to listen and respond
c) Is able to face an audience with confidence

These abilities are to be developed through exercises, which involve giving directions to reproduce a design, Or to reach a destination. Listen to announcements and follow directions. All of this is needless to say carried out in the English language.

The next focus is on critical thinking. This involves identifying and analyzing problems in difficult situations and making a justifiable evaluation. Holding debates on current issues of social or political relevance does this. Sometimes 'isms' become personal and at that point teacher intervention is called for.

From hereon the move is towards teamwork. Small projects within the campus are undertaken. Tying up with the N.S.S.UNIT to go for out of campus projects requiring group work is also encouraged. Planning a project, distributing work, executing individual tasks and being accountable for each job carried out are some of the points, which are evaluated. The ability to gather relevant information and manage it is also given a lot of importance.

In the third year entrepreneurship is introduced Students are given a chance to work out their novel ideas in the field of engineering at the TBI center. They are encouraged to identify the requirements of the industry and cater to it.

The skills that I have described above require either a teacher working solely towards this goal or extra efforts by those handling various subjects. The latter is too much of a strain and inevitably comes to a stand still. The former is more viable. Self-financing institutions have the ability to look at the finances involved and appoint necessary skill developers on a full time basis.

In the second year itself efforts are being made to introduce students to the industry by creating some tie-ups with chosen companies and the like to highlight the following requisites:

1. Manufacturing Methodology
2. Commercial Environment
3. Business Management Techniques
4. Manufacturing Process

In the third year some sort of paid project is to be taken up within or without the campus, which will serve to give some sort of work placement experience. The goal here is to:

1. Create commercial awareness
2. Enhance communication skills
3. Put team work into practice
4. Create self-confidence
5. Build up self-esteem

This is a new venture taken up by Jyothi Engineering College and we are optimistic about the outcome. To conclude, let me bring to your notice what NAASCOM (National Association of Software and Service Companies) has come out with. a study which states that some of the world's most admired and respected companies like Accenture, Convergys, DELL, Google, HP,IBM,ICICI,ITC,UTI Bank, Centurian Bank, Microsoft, Satyam and many more across the industry are in need of industry ready candidates

We have to exploit this opportunity and get our students placed with the readiness required of them. Necessary changes in curriculum and teaching methodology have to be looked at. The real challenge before the teaching community is positioning us as neutral ambassadors in providing inputs that are required through scientifically designed curricula and assessment to meet the demands of the industry today.

12

Emerging Job Opportunities for Fresh Engineers

*K.V. Varkey

ABSTRACT

The skill set requirements of engineers keep on changing in tandem with the advancements in technologies. Unless the engineering curriculum is not modified from time to time, the degrees and diplomas being conferred now will become obsolete. One has to be abreast with the changing technologies, to design "future proof" engineering curriculum. This paper highlights on some of the new technological developments in selected sectors that may result in the generation of new employment avenues for fresh engineering graduates and diploma holders and the challenges ahead. New employment opportunities and experiences in the sectors like, Energy, oil industry, Nano technology, IT, etc..., their dynamism in the nature of job market existing in these particular sectors are also analysed. The paper concludes that the experts of each domain can throw more light, to design future proof courses in promising disciplines. The speed of change in technology is increasing exponentially, thanks to the speedy research works made possible

*K.V.Varkey Pattimattam, Chief General Manager ,Hindustan Organic Chemicals Ltd, Kochi

by the information technology. Hence, unlike in the past, Universities cannot equip their present engineers to undertake the technological responsibilities for 30 to 40 years, ahead. Hence employers, in collaboration with universities have to make it a practice to impart continuing education for their engineers in latest technologies, so that their services can be utilised more effectively, for a longer period.

Introduction

The skill set requirements of engineers keep on changing in tandem with the advancements in technologies. Unless the engineering curriculum is not modified from time to time, the degrees and diplomas being conferred now, will become obsolete. One has to be abreast with the changing technologies, to design future proof engineering curriculum. Let us glance through some of the new developments in business/technology, which may result in the death of some of the existing job opportunities or generation of new employment avenues for fresh engineering graduates and diploma holders.

Jobs in Renewable Energy Sector

World wide renewable Energy Sector is growing thanks to the pressing need for reducing carbon dioxide levels in the atmosphere. In UK, 15% of the energy consumption will be generated from renewable sources by 2020 as per the Green Britain report. This requires the services of half a million people in the sector out of which 55,000 will be for wind energy ventures. In order to cater to this requirement, courses on renewable energy, domain are being introduced in various UK Universities.

Oil Sands and Shale Deposits

The spiralling cost of conventional oil has made the unconventional oil sources such as oil sands and shale gases, commercially viable. Oil sand mining in North Canada, is providing huge returns. Shale gas deposits are being exploited world wide, especially in USA and Canada. At present 15% of gas supplied in the US is sourced from shale gas deposits. M/s. Shell Chemicals have shale gas operations in the Netherlands, Singapore and USA. The Indian oil giant M/s. Reliance Industries also have ventured into the sector.

De-salination Plants

Drinking water will be a scarce commodity in the future. Hence re-cycling and re-use of water is gaining prominence. As per the cost dynamics, desalination is emerging as the most viable alternative. Many job opportunities are available in the huge desalination plants installed in Saudi Arabia, USA, Singapore, Australia, Israel etc. More such plants are being installed across the globe.

Green Building Technologies

Innovative developments are taking place in the construction sector, all over the world. Enhanced awareness on environmental hazards has led to energy efficient low cost building technologies and materials. Efforts are on, to build power generating buildings. The capability of Nano particles to act as photo -voltaic Cells is being utilized to meet this goal.

Dye sensitive Nano paint consisting of photo electric cells are printed/painted on steel surfaces to generate power.The thickness of the coating will be as low as 100 microns. M/s. Tata Steel is also among the steel giants venturing into the development of these kinds of innovative steels.

Intelligent buildings and roads with networked electronic devices will be popular in future.

Nano Technology

Nano technology will be redesigning many of the existing technologies with its ability to formulate materials of any desired properties. Bucky Balls, Graphene, Carbon nano Tubes etc will be utilized for the manufacturing of new products with unimaginable, capabilities. Curtains and walls/roofs coated with nano paints will be generating power for domestic use. Inrekor, UK has developed car chasis using laminated nano plastics which is as light as plastic but stronger than steel. Courses on nano sciences are introduced in many universities.

Information Technology - Today and Tomorrow

Right now, IT is offering millions of job opportunities for engineers. In addition to the conventional IT jobs like software development, net working, sales and service etc, there are so many emerging areas such as chemo informatics, bio informatics, mechatronics , Enterprise

Resource Planning, Local language computing etc. IT based jobs have become accessible to engineers from all the branches as IT is known to all engineers just like English is there with all graduates.

The billing rates of software development jobs are coming down due to cut throat competition.Some out sourcing customers are adopting reverse auction, resulting in cost saving upto 40%. As per the recently published study of ASSOCHAM, India is losing its edge to countries like Philippines in BPO jobs also.

In order to maintain Indian domination in this industry, the following strategies are to be adopted

a) Concentrate on emerging hi-tech verticals
b) Open technology parks in low cost rural centres
c) Short-term skill development courses for graduates/diploma holders in Engg.

Global Certifications

Students from developing countries like India face lack of accreditation of their Universities abroad. Major technology leaders have set up a level playing field for the students all over the world by introducing global certification examinations in specialised areas. By acquiring these qualifications they get employed irrespective of their country of origin. These examinations are held on-line and can be taken up by anybody. But it will be easier for the qualified engineers to get through them. A few examples of such courses are given below:

a) MCSE,MCDBA,MCSD,MCT,MCBC,MCPD,MOS etc by Microsoft.
b) Linux certification courses by Red Hat, SAIR and Linux Professional Institute.
c) CCIE, CCNP,CCNA,CCDP,CCDA,CCVP,CCIP, CISSP etc – by Cisco.
d) Computer servicing certification course (A+), Networking course (N+)and Security Course (security +) – by Comp TIA.

Knowledge Process Outsourcing

Knowledge Process Outsourcing (KPO) sector is a gold mine of engineering jobs. In the BPO sector. engineers are required only for the technology management. whereas, in KPO , highly qualified engineers are required for specialized ground level jobs like, design, research, legal services etc. The salary levels in KPO will be much higher than BPO.

Business Intelligence Technologies

Progressive organizations are employing business intelligence technologies on top of the ERP software deployed for taking informed decisions in time. Training in packages like, Cognos, Terradata, Hyperion etc will be helpful to get employed in this sector.

Knowledge Management

Of late, knowledge is considered as an asset and there are openings for knowledge engineers, knowledge navigators, knowledge managers etc in the corporates. Training in packages like e-gain, knowledge navigation etc. helps engineers to enter this field.

Cyber Forensics & IT Security

In order to fight the menace of cyber criminals, lot of money is spent on IT security and cyber forensics Smart qualified engineers are required in millions for ethical hacking, security consultancy etc.

Personal Robots

As per the study of the International Federation of Robots, this sector can create one million high quality jobs in the world in next 5 years. Opportunities will be available in electronics, semi conductors, pharma, ship building etc. As Robots can ensure high quality, speed, precision and traceability its role in industry is increasing day by day. Personal Robots will be doing the domestic work in future homes. M/s. Willow Garage have already introduced a domestic robot which works on Robotic Operating System.

Gaming

As per the NASSCOM report, gaming industry will be growing at an annual rate of up to 53%. Gaming will find jobs both in India and abroad. There are opportunities for programmers, developers, designers, artists, testers and producers. Graduate engineers and diploma holders can grab the opportunities through short term courses on Shockwave, J2ME,Open GL, Flash, Maya, 3DS Max, Mudbox, Xara etc.

Outsourced Research Works

As an outcome of globalisation, many multinational firms are outsourcing their research activities to the locations where qualified manpower is available at low cost. The R&D centres mushrooming in Bangalooru, Noida, Pune etc in sectors like software, pharamaceutical, automobiles etc are examples. Opportunities are expected in many inter disciplinary and cross-functional research activities. Post graduate/ doctorate level education is to be encouraged in selected areas, in order to capture this job market.

Consultancy

Other than the conventional consultancy jobs, many new avenues are being opened for engineers. Qualified engineers are required for advanced information technologies,intellectual property rights, patenting, carbon credits etc. Institutions like IIT Kharagpur have commenced post graduate courses in intellectual property disciplines.

Other Areas

In the knowledge era new job opportunities are christened day by day. Digital surveillance, digitatisation of books and documents, digital arts, wireless technologies , embedded system technologies etc. are few examples. Jobs related to environmental engineering, safety and logistics are also on the rise. Conventional domains of manufacturing and construction will be offering millions of job opportunities at any point of time.

Conclusion

What we have seen is only a few examples of the dynamic nature of job market. Experts of each domain can throw more light, to design future proof courses in promising disciplines. The speed of change in technology is increasing exponentially, thanks to the speedy research works made possible by the information technology. Hence, unlike in the past, Universities cannot equip their present engineers to undertake the technological responsibilities for 30 to 40 years, ahead. Hence employers, in collaboration with universities have to make it a practice to impart continuing education for their engineers in latest technologies, so that their services can be utilised more effectively, for a longer period.

❑❑❑

13

Profile of Technical Education Facilities

Employment Scenario in Karnataka State

Dr. B.M. Dodamani and Aruna Kumar Shetty B.

ABSTRACT

NTMIS Nodal Centre for Karnataka state at NITK, Surathkal, has been regularly collecting data and maintaining an up-to-date data bank in order to provide reliable information on professional institutions like Engineering, Management, Hotel management & catering technology & Pharmacy and also on the passed out students from these institutions. An attempt has been made in the present paper to analyze the facilities available for the expansion and development of technical education in engineering degree and diploma level. This paper focuses on some of the vital parameters related to technical education such as growth of institutions, sanctioned intake, actual admission, and for the year 2005 to 2009 in Karnataka and teacher-student ratio, employment profiles such as first paid activity and their paid salary for the degree and diploma engineers for the Karnataka state during the batch year 2009.

**Dr. B.M. Dodamani and Aruna Kumar Shetty B*, NTMIS Nodal Centre, National Institute of Technology Karnataka, Surathkal

Introduction

Karnataka State finds a prominent place on the map of technical education in India after independence of the country in view of the large number of technical institutions offering variety of quality education to its stakeholders. It stands at 4th position, next to Tamil Nadu, Maharashtra and Andhra Pradesh in terms of technical institutions in the state. An attempt has been made in this report to trace the facilities available for technical education upto the year 2009 in the state.

Technical Education in Karnataka State has been rapidly growing for the last 3 decades and there has been a conspicuous expansion in number of institutions imparting technical education in terms of quality and quantity offering engineering degree level courses and also in emerging and specialized fields. In view of the fact the state government provided more inputs to make the technical education a popular front for development of the nation, it also had in mind the concept of social justice and concern for downtrodden. Private institution is pioneer in Karnataka State.

Growth of Institutions

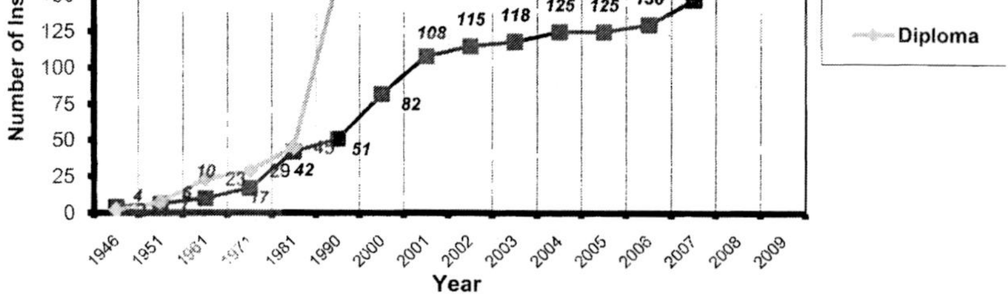

The Growth of Technical Education Facilities for degree and diploma courses in Karnataka between the years 1946 to 2009 is illustrated in Figure 1. During the year 1946, 4 engineering and 2 diploma colleges existed and during the year 2000, 82 engineering colleges and 185 diploma colleges were functioning in the state. 93

engineering colleges were sanctioned during the year 2001-2009. At the end of the year 2009, there were 175 engineering colleges and 274 diploma colleges existing in Karnataka State. This increased demand for technical education reflects the impact of post liberalization economic growth and the resultant higher employment prospects.

Sanctioned Intake

The year-wise sanctioned intake into engineering degree and diploma courses for the reference period 2005-2009, is presented in Table 1 below. The sanctioned intake at degree level in the year 2005 was 46,214 and it increased to 74,322 during the year 2009. The variation increase is nearly 61 percent from the year 2005 to 2009.

The year-wise sanctioned intake into diploma courses in the year 2005 was 34,343. The intake increased to 76,104 in the year 2009 recording an increase of more than 125 per cent during the period of next 5 years from 2005 to 2009.

Table 1 : Year-wise Sanctioned Intake into Engineering Courses from 2005-2009

Sl.	Level	2005	2006	2007	2008	2009
1	Engineering Degree	46214	48079	52559	56479	74322
2	Engineering Diploma	34343	36322	38138	38198	76104

It may be observed that the sanctioned intake in respect of engineering degree and diploma courses are significant. This could be mainly due to rapidly changing technological scenario and significant shift in the global labour market environment owing to changed pattern of trade, political & economic concerns, competitiveness, innovations in information technology, etc.

Actual Admission

The actual admission into degree courses and diploma courses is presented in Table 2. The actual intake into degree level courses during the year 2009 was 56,358 against the sanctioned intake of 74,322. The admission was nearly 76 per cent with respect to the sanctioned intake.

Table 2 : Year-wise Actual Admission into Engineering Courses from 2005 - 2009

Sl.	Level	2005	2006	2007	2008	2009
1	Engineering Degree	45011	45403	47364	50828	56358
2	Engineering Diploma	28806	29024	30510	30250	57453

The actual admissions made into diploma level during the year 2009-10 were 57,453 against the sanctioned intake of 76,074. The actual admission into diploma courses was more than 76 per cent with respect to the sanctioned intake.

Outturn

Table 3 shows the year wise data on outturn of degree and diploma students for the years from 2005-2009. The total outturn at degree level for the year 2009 was 40,089. The total outturn of degree students during the year 2005 was 29,615. Similarly, the outturn increased to 40,089 in the year 2009 after showing an increase of another 35 per cent in the later five years (from 2005-2009).

Table 3 : Year-wise Outturn of Engineering holders from 2005-2009

Sl.	Level	2005	2006	2007	2008	2009
1	Engineering Degree	29615	27848	28641	33672	40089
2	Engineering Diploma	12075	11919	11845	16306	13274

The year wise data of outturn of diploma students depicts that the increase in outturn is marginal. The percentage variation was 10 per cent for the period from 2005 to 2009. It can be concluded that the increase in outturn for the period from 2005 to 2009 was gradual at diploma level. Total pass rates for the total students appeared for exam are less at diploma while comparing with the engineering degree courses.

Teacher-student Ratio

The data pertaining to number of teachers available and the enrollment of students from first to final year in engineering degree institutions are exhibited in Table 4. It is revealed that 16,122 number of teachers were available in the year 2009 for the total enrollment of 2,28,324 students at degree level. The average teacher-student ratio was 1:14. In many a departments, the number of students per teacher was good apparently due to the fact that teachers of other departments shared the workload of such departments.

Table 4 : Teacher-Student Ratio in Engineering Colleges by Level in 2009

Sl.	Level	No. of Teachers	Student Enrolment	Teacher-Student Ratio
1	Engineering Degree	16122	228324	1 : 14
2	Engineering Diploma	7011	89026	1 : 13

The data pertaining to number of teachers available and the enrollment of students from first to final year in diploma institutions are also exhibited in Table 4. It is revealed that there were 7,011 teachers available in the year 2009 for total enrolment of 89,026 students at diploma level. The overall teacher-student ratio was 1:13. In many departments, the number of student per teacher was good probably due to teachers of other departments shared the workload of such departments.

Employment - Activity status

The information on activity status was collected through the graduate follow-up questionnaire from the students passed during the year 2009. The reference date for collection of information was 31st May, 2010, i.e., leaving 12 months gap after the announcement of results. The various types of activities considered in this study are: Paid Employment (India and Abroad), Unemployment Status, Higher Studies and other Activity (Self Employed in Family or Own Enterprises, Apprenticeship Training, Not Looking for Job, Interested in Self-Employment).

The activity status of engineering degree and diploma holders for the outturn batch year 2009 is presented in Table 5. It is seen that nearly 45% degree holders are engaged in paid job (in India and abroad) after passing their degree programme. Nearly 32% are unemployed and are looking for a job. It is encouraging to notice that

18% were pursuing higher studies. Thus, it is observed that the activity status were mainly distributed among paid job (45%), unemployed and looking for job (32%) & involved in studying (18%). The remaining percentages of degree holders were involved in other activities. The activity status indicates that even after waiting of 12 months, unemployment of the degree level has not improved much. In conclusion it can be inferred that degree holders engaged in paid job is relatively less while comparing with the unemployed engineers looking for job. However, it is an assuring development that more percentage of engineers are pursuing higher education.

The activity status of engineering diploma holders for the outturn batch year 2009 is also presented in Table 5. It is seen that 20% Diploma holders are engaged in paid job after passing their diploma programme. It is interesting and encouraging to notice that almost 44% diploma holders pursued higher studies after passing out from the polytechnics.

Table 5 : Activity Status of Engineering Holders by Level - 2009 batch

Sl.	Level	Activity Status as on 30th May, 2010					
		Paid Employment India	*Paid Employment* Abroad	*Unemployed & Looking for Job*	*Studying*	*Others*	*Total*
1	Engineering Degree	17895	111	12907	7287	1849	40059
	%	*44.7*	*0.3*	*32.2*	*18.2*	*4.6*	*100*
2	Engineering Diploma	2592	0	3813	5823	972	13200
	%	*19.6*	*0.0*	*28.9*	*44.1*	*7.4*	*100*

Average Emoluments

The average monthly emoluments earned by engineering graduates in their first paid employment are shown in Table 6. The monthly earnings received by the graduates are reckoned in 3 ways. i.e; Maximum emoluments received; Minimum emoluments received and Weighted average emoluments. The overall weighted average of salary received by the degree holders was '16,851. The maximum overall average salary was '46,848 and minimum overall average salary was '6,798. There was a gap of nearly 7 times between minimum average salary and maximum average emoluments. The trend in first paid emoluments in India and Abroad depicts an increase among degree engineers during the reference year 2009 compared to year 2008. It is noted that the earnings differentials were wide among various disciplines.

Table 6 : Average Monthly Emoluments of Engineers by level - 2009 batch

Sl.	Level	Emoluments (in ')		
		Maximum	Minimum	Weighted Average
1	Engineering Degree	46848.00	6798.00	16851.00
2	Engineering Diploma	15605.00	3400.00	7415.00

The average monthly emoluments earned by engineering diploma in their first paid employment are also shown in Table 6. The monthly earnings received by the graduates are reckoned in 3 ways. i.e; Maximum emoluments received; Minimum emoluments received and Weighted average emoluments. The overall weighted average of salary received by the degree holders was '7,415. The maximum overall average salary was '15,605 and minimum overall average salary was '3,400. There was a gap of nearly 5 times between minimum average salary and maximum average emoluments. The trend in first paid emoluments in India depicts an increase among diploma engineers during the reference year 2009 compared to year 2008. It is noted that the earnings differentials were wide among various disciplines.

Conclusions

The institutional information collected for the current reference year 2008-2009 in respect of degree and diploma courses evinces that the technical education is consistently expanding in Karnataka State. This expansion reflects the steady economic growth in terms of industrial expansion, technical skill developments, storage of information systems, R&D activities and so on. The impact of post liberalization economic activities resulted and reflected in the form of augmented employment prospects. The contributions of the private investors are the main reason for the greater expansion of technical education in Karnataka State. However, lack of incentives to private players resulted in decline of admission to the extent of sanctioned intake of private institutions in recent years. The outturn figures envisaged better prospects for degree courses. Since quality of education is determined by the availability of qualified teachers at institutions, it is mandatory to provide ways and means for teachers to acquire higher qualifications under various schemes. The teacher-student ratio is at an encouraging level. Hence, it is noticed from the collected data that the current arrangements of teaching facilities in degree level institutions are satisfactory in Karnataka State.

Further, the number of institutions, sanctioned intake and actual admissions show that the preference for diploma holders was comparatively less. This indirectly suggests us to extend more support for diploma courses. The contributions of the private investors are the main reason for the greater expansion of technical education in Karnataka State. Since quality of education is determined by the availability of qualified teachers at institutions, it is mandatory to provide ways and means for teachers to acquire higher qualifications under various schemes. It can be inferred as a whole that the diploma education needs further attention to gear-up to achieve the anticipated goal.

The student follow-up survey conducted for the reference batch year 2009 suggests that the percentage of paid employment was encouraging among degree holders. While examining the unemployed degree holders seeking employment, considerably large (32%) percentage of engineers were represented in this category. The total engineers, pursuing higher studies were 19% and this percentage is slightly better while comparing with year 2008. The maximum average emoluments ('46,848) were almost 6.5 times higher than the minimum ('6,798) average emoluments in the first paid employment at degree level. The student follow-up survey conducted for the reference year 2009 suggests that none of the diploma holders obtained employment abroad. The total diploma engineers who pursued higher studies were encouragingly more (29%), while comparing with degree holders.

Reference

1. Annual Technical Manpower Review Karnataka - 2009.

14

Employability of Fresh Engineering Graduates:

Problems and Prospects

*Murali Pasupathy

ABSTRACT

Today there are more than 2300 Engineering Institutions in the country turning out more than 600,000 engineers every year. Indian Engineering Education represents one of the largest educational systems in the world. The challenges posed by this rapidly growing system in our country are extremely complex. As we address the needs of a growing knowledge economy in India, quality of Engineering Education emerges as an issue of paramount importance.

Engineering education is not relevant for a graduate if it does not meet the needs of the industry. A core set of such needs must be identified. It should include skills related to manufacturing, computer technology, interpersonal relations, analytical and problem solving techniques. Academic programs have to be evaluated and revised periodically so that they endow graduates

*Chief Operating Officer, Naethra Technologies Private Limited, Chennai *with skills pertaining to the above mentioned set. The faculty should also be equipped with proper teaching skills and aids to impart quality education to the students.*

Having mentioned this, the focus should be drawn on training the mind and developing the power of analysis. This can be accomplished through a proper and well balanced curriculum and appropriate grading system. It should be remembered that the primary philosophy of teaching should be to collect, collate and feed the information to the students in a digestible form. The students should also be assessed on a continuous basis to test the acquired skills.

The engineering graduates of the advanced countries possess a high level of confidence in handling unknown problems. Many of our graduates are deficient in handling unknown problems. The gap has to be bridged through setting up appropriate educational system.

This presentation deals with the problems faced by the Engineering Education system in India and some of the possible remedies that can be implemented to rectify these issues.

Introduction

We often hear two contrasting views of the employment scenario. On the one hand, employers cry hoarse about non-availability of quality talent in the market, while on the other, we hear of millions of unemployed youth, who are unable to find a well paid job several years after completing their engineering or professional degrees. A recent survey suggests that more than 70% of the engineering graduates in India are unemployable. What ails these graduates or the so called degree holders? Why are they not suitable for employment?

Key Issues

Some of the key issues that fresh graduates face while seeking employment:

- Inadequate exposure to practical aspects
- Inadequate English communication skills – Oral and Written
- Faculty have been completely insulated from the industry
- Very minimal spent on up skilling of the faculty members

- Ineffective problem recognition and solving skills
- Lack of business exposure
- Innovative thinking is not encouraged at the student level
- Lack of understanding of opportunities in the Indian market
- Brain drain – is still a major issue despite a reduction in exodus

Global Challenges

More and more companies globalize and look opportunities across the world. So, engineers face additional challenges to keep themselves in their jobs:

- Discipline and specialization oriented vs multi and cross disciplinary endeavors
- Need for professional and business skills
- Globalization of economic activities
- Mobility of professionals across national borders
- Ability to work harmoniously and effectively in multicultural teams
- Rapid obsolescence of technology
- Need for life-long learning attitude and self learning abilities
- Impact of global warming on existing technologies

Suggestions for Improvement

- Set up laboratories similar to Smithsonian Labs in the US
- Real life issues should be taken up for solutioning at colleges (For e.g.: They can visit TPM practicing companies)
- Training on system based lecture delivery and e-learning
- Students can be given small Mobile app development projects. These projects require very minimal budgets and can be built in a short time span
- Data collection projects have to be allocated to students
- Student exchange programs with global universities
- Students can contribute to technical journals to improve written communication skills

Some of the sample projects that can be allocated for engineering students:

Discipline	Sample Project
Computer Science & Engineering	Develop small web sites for businesses or build an app to publish semester results on a mobile phone
Electrical & Electronics	Design a circuit diagram for a house, small factory. Variations could be to build a circuit diagram for Solar/ Wind power generation & test a prototype
Electronics & Communication Engineering	Design a Water level controller or join a HAM club
Civil Engineering	Design a plan/elevation for a simple structure in the college premises & get it approved by an architect
Mechanical Engineering	Take an IC engine apart and understand how it functions
Industrial Engineering	Take up factory visits or Work on a market survey & interpret results

Engineering Education - Changing Scenario

- Merging of disciplines
- Combine domain knowledge with sound appreciation and familiarity with interdisciplinary projects
- Ability to recognize problems of integration of their domain activities with those of engineers of other disciplines
- New Learning Paradigms
- Student – centric learning, teacher acts as a facilitator, guest lecturers from Industry

- Project based learning
- Case study methods and design simulations
- Concern for Environment and Sustainability
- Stretching of natural resources
- Energy efficiency in focus
- Ability to understand the impact on environment and ecology
- Alternative Delivery Modes
- Rapid advancement in Information and Communications Technology
- Face To Face, Distance Education, Online etc
- Need for strict Quality Control mechanisms

Engineering Education - Faculty Shortage

- Total shortage of qualified engineering faculty - 1,30,000
- Desired educational qualification is at doctoral level
- Most of the institutes are not able to appoint faculty members with Masters Degrees
- Faculty members do not have knowledge of teaching pedagogy
- Teaching community is inadequately positioned to connect with the students
- Lack of adequate compensation is a huge challenge
- Quality Improvement Program (QIP) has yielded desired results across multiple institutions

Engineering Education - Curriculum Changes

- Engineering education should equip people with new and relevant skills to kindle new ideas
- Engineering practices based on empirical studies and data alone may not give us the cutting edge
- Examinations have to test the concepts and understanding of the subject
- Curriculum should include minimum requirements for humanities, science, core engineering subjects, specialized electives, foreign languages etc
- Social awareness, economic developments, business processes and money management in business have to form part of curricula

Engineering Education – Role of Ethics

- Dictionary definition of ethics – It is a science of morals, or a set of principles with a sense of purpose
- Ethics are fundamental to civil society
- Ethics have been adopted as an integral part of teaching-learning process
- Engineers are expected to apply science and technology for the progress and prosperity of society at large
- Engineers interface with society in multiple modes:
- Professional worker
- Teacher
- Manager

Global Engineering Concepts – Road Blocks for India

- Preparation for global practice is not central to an engineer's education
- International mobility remains a challenge
- Key challenges - visas, costs, cultural barriers, language and curriculum
- Policy decision of permitting foreign engineering education providers
- Meaningful and substantive partnerships with global institutions
- Lack of knowledge about proven theories and effective practices to instill global competence
- Programs lack rigorous and scientific evaluation needed to assess the educational impact of proven theories

Conclusion

In addition to unemployment, we are bringing in a kind of under employment for engineers where people do work that is much below the skill sets that they are supposed to have. These resources will not be able to go up the corporate ladder and it will be very difficult to give them a career path. Ideally, we would want engineering graduates to up skill themselves and move up the value chain so that they don't become redundant. The government and the society at large has to wake up to the reality that one day an African country might become the world's back office and IT hub and do to us what Bangalore or Chennai did to the western world. Before that, we have to make sure our education system gets upgraded suitably and can mass produce well rounded personalities who can take India to the path of glory.

□□□

15

Engineering Labour Market in Kerala

Quantity and Quality Issues

Dhanya Menon and Zakkariya K.A.***

ABSTRACT

The Indian labour market has been witnessing a dearth of employable engineers even though the number of engineering colleges have increased manifold since independence. Nevertheless Kerala is considered to be home to the most employable IT engineers compared to other southern states but the situation here too is far from dismal. High dropout and failure rates at the undergraduate level and high unemployment among the graduating engineers continues to be grave issues. Among the glut of reasons for the low employable output like government sanctioning of engineering colleges without regard to demand factor, operation of institutions lacking in infrastructure and academic necessities and relaxed eligibility criterion for admission, the one reiterating issue has been the deteriorating supply of qualified faculty. Policy measures to resolve the demand supply gap have been focussing on the

* Research Scholar, School of Management Studies, Cochin University of Science and Technology

** Reader, School of Management Studies, Cochin University of Science and Technology

access, equity and quality issues but the results have been discouraging. It should be emphasised that there is a lack of effort at the institutional level to answer the issue. Promotion of academic excellence and improvement of employability of student output should be among the top priorities of engineering colleges to ensure a good supply of labour to satisfy the industry demand of our burgeoning economy.

Introduction

The Indian labour market has been witnessing a dearth of engineers since independence even though the number of educational institutions offering engineering courses has increased since Independence. Adding to the woe of lack of engineering graduates is the surmounting lack of employability among the available engineers. Thus, we have the quantity and quality issues in the engineering labour market- the quantity issue of lack of engineering graduates due to increased drop out and student failure and the quality issue due to lack of employability of the graduating engineers. The 2011 National Employability Survey conducted by Aspiring Minds found the state of Kerala leading in the employability of its engineering graduates in the IT sector but the situation in Kerala too is far from dismal.

Evolution of Engineering Education

Engineering education was introduced in India by the British mainly to cater to their labour needs in civil and construction works. Some of the prominent engineering colleges set up in India in the pre-independence era that continue to exist till today are the Indian Institute of Technology, Roorkee (1847), College of Engineering, Pune (1854), Bengal Engineering and Science University, Shibpur (1856), Banaras Hindu University (1916) and Harcourt Butler Technological Institute, Kanpur (1920). The first engineering college in Kerala, College of Engineering, Trivandrum was established in the year 1940.

India's top tier engineering institutions the IITs were the result of the Government's initiative in advancing technical education. The initiative led way to the foremost IITs in India at Kharagpur (1950), Bombay (1958), Kanpur (1959), Madras (1960) and Delhi (1961). It paid off as the institutions continue to be the role model for regional institutions and also are the only representation of India in global rankings though not in the top 100.

The Kerala state government however, waited nearly two decades to set up a second one thus witnessing no growth in the number of engineering colleges during the period 1940-60. It saw moderate growth during 1960 to 1980, normal growth in 1980-98 periods and a high growth witnessed only after 1999 as the state permitted self finance colleges in the private sector.

According to AICTE statistics as of 2011-12 Kerala has 177 institutes offering engineering andtechnology courses out of which 66 are government institutions, 14 governments aided, 86 are unaided private institutions and 11 are university managed institutions.

Student Outturn and Unemployment

A newspaper report by Kumar (2007) dwells succinctly into the engineering student outturn of the state. A study done in Kerala had shown that even when there was only less than 5,000 seats for engineering, about 10 per cent of the students were unable to complete the course successfully. About 90 per cent of the incoming students used to turn out at the end of the fourth year, during the 80s. Down the years the situation had not improved with the outturn in 2003 being only 58 per cent of the intake of 1999, which was around 6,700. The survey conducted by the National Technical Manpower Information Systems had found that the outturn of engineering graduates in Kerala was only 3,944 against the actual intake of 16,606 students in 2003-2004 which led to official estimates on the outturn in 2008 to be 50 per cent less than the actual intake.

The sudden spurt into the student outturn statistics was a result of the declining quality of engineering education hitting headlines when the final year B. Tech results of Cochin University of Science and Technology were announced in 2007. Out of the 3,100 students who wrote the final semester examination only 1,023 had passed – a meagre 33.02 per cent. The number of first class with distinction holders was only 346. More shocking revelations came to the limelight when it was found that only 4 candidates passed the examination out of the 295 supplementary candidates (1999-2002). The varsity was finding it difficult to ensure excellence with a steady increase in the number of engineering colleges.

But all is not well even for the lucky few who pass out for to make matters worse, out of the numbered graduates that emerge every year only some of them get placed through campus recruitment, a few manage to find a job on their own and a tiny few choose to be self employed. Though enrolling for higher studies is not something to be looked down upon it is but hard not to do so these days as it has become a means for youngsters to disguise their apparent unemployment. Those of whom who don't make it to either employment or higher studies are left with the option of travelling if not to the mercy of unemployment subsidies.

Table 1 : Outturn and Unemployment Rates of Engineering Graduates in Kerala

Year	Sanctioned Intake	Actual Intake	Outturn	Work Seekers
1999	6668	6126	3531	8334
2000	8820	8739	3836	9858
2001	11293	11147	4089	10349
2002	18280	16143	3724	8506
2003	19889	16563	3944	8116
2004	23643	16810	6639	9661

Source: *Annual Technical Manpower Review (Engineering) Kerala and Directorate of Employment and Training, Kerala*

A comparison between the figures released by the Annual Technical Manpower Review and the Directorate of Employment and Training of Kerala will show that the low number of graduating engineering students when they enter the labour market have to compete with the existing unemployed engineering graduates to win a job they both aspire failure in which will result in them adding to the figure of the growing number of engineering graduate work seekers. The only upper hand they may have will be that they are fresh from school. Also, the disproportionate increase in the number of engineering graduates each year compared to the increased number ofengineering colleges is but a reflection of the stagnant quality levels of the institutions. A huge percentage of the engineering students that finally do graduate often find themselves jobless and left to fend for themselves in the market. The irony that they are not being absorbed into the industry as such viewing the dearth of supply of engineers as voiced by industry is but a graver issue.

Reasons for Demand Supply Gap

There are many facets to the demand gap scenario. Firstly although the number of engineering colleges and enrolments have increased it is but a miniscule of the Indian population who has access to higher education with not more than 7% of Indians aged 18-25 going to college. There is also a lack of equitable representation. For example, in a study conducted by the Centre for Social Exclusion and Inclusive Policy, CUSAT in 2011 social, economic and gender factors were found to be highly affecting the ranks obtained by students at engineering entrance examinations in Kerala. Moreover, as the national level and best institutes are generating too few graduates the regional and low tier institutions are producing graduates of uneven quality. The demand supply gap is a serious issue as lack of human capital to feed the industry demands can actually have a profound effect and slow down our economic progress.

Numerous studies have been conducted to determine the causes for the lack of quality technical education in India in addition to the periodic Government Committee reports on technical education. The absence of a need based study while sanctioning colleges and courses has been cited as the numerous reason for the demand supply mismatch. Students opting for popular engineering courses rather than the ones in demand by the industry has widened the demand supply gap.

Non availability of quality faculty to teach, a nation wide problem is a time and again cited one. A survey conducted in 2006 had found that the country was facing 65% shortage of qualified teachers and the appointment of BTech degree holders as teachers in engineering colleges was suggested as a short term measure. The problem arose when engineering colleges were increasingly found providing permanence to it. In Kerala the issue was recently reiterated in the inspection committee report by the Directorate of Technical Education, Kerala on 105 self financing engineering colleges. A surprise inspection was conducted to verify the physical availability of faculty in the colleges with regard to the declared strength for university affliation under the directive of the high court. The colleges, 13 under the MG University, four under the Kerala University, four under the CUSAT and two under Calicut University were found lacking in infrastructure and academic necessities. The colleges under the Kannur University were found to be meeting the minimum requirements and only five of the 24 colleges inspected had qualified teachers. The report stated that the salary, allowances and other welfare schemes available for the faculty members in majority of the colleges were pathetic. The committee recommended that only postgraduates with valid GATE score should be recruited to teaching posts in engineering colleges and that the government should take it up with the Kerala Public Service Commission to ensure it. The issue assumed significance as the Kerala government decided to approve the starting up of 19 new engineering colleges in 2011-2012. The college managements would have to fill many faculty posts in various engineering streams.

Another common reason cited for student failure is lack of selectivity. Although the Government has set some minimum performance standard in the common entrance test as an eligibility criterion over and above the minimum marks specified for the qualifying examination it is held that it is sooner or later relaxed under pressure from self financing college managements.

Those of whom who do slog it out at the engineering colleges find it hard to land on a good job offer for lack of employability. According to a survey, jointly carried out by the Federation of Indian Chambers of Commerce and Industry (FICCI) and the World Bank in 2009, 64 percent of surveyed employers are "somewhat", "not very", or "not at all" satisfied with the quality of engineering graduates' skills. The top three most important general skills identified were integrity, reliability and teamwork, while the top three most important specific skills were entrepreneurship, communication in English and use of modern tools and technologies. The survey has said that employers are relatively satisfied with the graduates when it comes to communication skills in English, but not with the graduates' reliability. Lack of interaction with industry in designing education programs, lack of definition of set of skills to be acquired by graduate after each semester and teacher centric pedagogical style were cited as main reasons for lack of employable skills of engineering graduates.

Policy Measures

The current policy measures to improve the supply and quality of engineering graduates mainly concentrate on encouraging the participation of private sector, performance evaluation and introduction of quality sustenance procedures in universities and colleges by accreditation bodiessuch as NAAC and introduction of Quality Improvement Programs for faculty.

Other than the measures for higher industry collaboration in matters such as development of industry specific curriculum, twinning research programs etc and the option for students to join finishing schools after graduating, few measures have been taken seriously to improve employability of graduating students. NAC-Tech (NASSCOM Assessment of Competence- Technology) conducted by NASSCOM are among the few available industry standard assessment and certification program for final year and pre-final year engineering students on the employability front with the scores of all NAC-Tech test takers being shared primarily with all the endorsing companies enabling them to directly connect to those who perform well in the NACTech assessment and meet defined job criteria.

Lack of Institutional Level Effort

Institutional level initiatives to better the twin issues of quality and quantity of engineering graduates have been found successful by many researchers. Though the fact has been widely acknowledged very few engineering colleges are implementing them. Kaushik and Khandujah (2010) had studied the implications of applying Six Sigma methodology over a technical institute to increase the passing rate of students and found that application of Six Sigma project recommendation increased the sigma level of the technical institute to 4.17 from 2.28 and significantly resulted in increasing the passing rate of students. The industry concept of reducing defects was successfully transplanted to the education scenario's defect alias unsuccessful students.

Promoting Academic Excellence and Improving Employability of Students

The domain of academic excellence and industry preference for recruitment is sadly restricted to the IITs in India and the fact that the top national institutions fail to figure up in the top tier in international rankings is but an eye opener to the need for measures to improve the quality of technical education in India. Vrat (2009) had identified indifference, lack of passion/zeal, lack of commitment to organizational objectives, lack of high performance teams and negative attitude as the major excellence inhibitors. Thus, measures to overcome these contagious should be the top priority of the administration of educational institutions.

Conclusion

Quantity and quality issues of low student outturn and unemployability can be partly offset if the existing policy measures are supplemented by an institutional level effort which has been initiated successfully but only at a few pioneering institutions. Introduction of better management practices requires a student centric approach which is possible only with the active participation and cooperation of institutional heads and faculty. All the more there is an urgent need to address this issue not only at the state policy level but from the institutional level also for reasons more than economic.

References

1. D Rajasenan and Binu P Paul (2011): Technical Education: A Long Way to Go, Working Paper, Centre for Social Exclusion and Inclusive Policy, Cochin University of Science and Technology, Editorial dated 5/2/2011, Mathrubhumi
2. Directorate of Technical Education, Kerala (2011): Inspection Report on Engineering Colleges Under CUSAT, Report submitted to Honourable High Court of Kerala as per directions in the Judgment WA No. 2014 of 18-1-2011 and WA No. 1319 of 2011 dated 22/9/2011
3. G Krishna Kumar (2007): Wither Engineering Education?, Article dated 14/8/2007, Education Plus, The Hindu
4. Ganesh Hegde and V Umashankar (2011): Accredition by NAAC for the Third Cycle: Issues and Challenges, University News Special Issue on Reforms and Innovations in Higher Education: A March towards World Class Standards, Vol. 49 No. 50

5. Ministry of Human Resource Development, India (2001): Technical Education Quality Improvement Project of Government of India, Document of Department of Secondary and Higher Education
6. Muhammed Aslam MK and PR Wilson (2009): Problems and Prospects of Higher Education in Kerala, Technical Education in Kerala: Emerging Trends compiled by Meera Bai M, Serials Publications, New Delhi
7. Planning Commission, India (2007): Report of the Steering Committee on Secondary, Higher and Technical Education for the Eleventh Five Year Plan (2007-2012)
8. Prabhakar Kaushik and Dinesh Khanduja (2010): Utilising Six Sigma for Improving Pass Percentage of Students: A Technical Institute Case Study, Educational Research and Review, Vol. 5, No. 9, pp. 471-483
9. Prem Vrat (2009): Academic Excellence in Technical Education: The Role of Organization Leadership and Industry Collaboration, Paper presented at Workshop by Maharashtra Board of Technical Education at VIT, Pune
10. Rangan Banerjee and Vinayak P. Muley (2008): Engineering Education in India, Paper presented at seminar sponsored by Observer Research Foundation at IIT Bombay
11. World Bank (2000): Scientific and Technical Manpower Development in India
12. Zakkariya KA and Jagathy Raj VP (2009): An introspection to the Problems and Issues of Engineering Education in Kerala, Technical Education in Kerala: Emerging Trends compiled by Meera Bai M, Serials Publications, New Delhi

16

Employment Scenario of Fresh Degree and Diploma Holders
Problems of Unemployment among the Engineers in Kerala

**Shelly M.K.*

ABSTRACT

The paper highlights on the various dimensions of the problem of employment and unemployment among the technically qualified professionals in Kerala. The classification of the engineers into the diploma holders and graduate engineers and their employment status aims at identifying the basic difference between these groups with respect to their employment status and subgroups further categorized into gender, social class, type of institution of study, region of the location of the institution, the parental education of the students, etc. Thus various factors that influence the employment status of the students are tested in order to decipher the influence of exogenous factors in determining the employment of the qualified engineers. This holds importance on the background of the then prevailing economic recession during the year 2009 and therefore

*Shelly M. K., ICSSR Fellow, Department of Applied Economics, CUSAT

the extent to which the engineers of this period has been affected due to the global recession is addressed and expectedly more of the relatively undereducated bear the brunt of unemployment . The incidence of larger proportion of unemployment among the professionals leads to its disrepute and decline in the faith on the professional course as a guarantor of employment and therefore the educational planning of the nation suffers a major blow .The extent to which this gets affected is known only if the actual status of the professionally qualified personnels are unveiled by detailed examination of the facts. Thus the paper addresses on the scope of engineering education in Kerala in the years to come.

Introduction

The educational system in Kerala comprises of the different stages with primary, secondary and senior secondary level imparted through general education and higher education that consist of undergraduate and postgraduate level including the professional courses with Technical education being its integral part. But in recent times the drastic changes through wider expansion of the higher education level is manifested due to the active involvement of the private sector in this area. The growing scope of professional education imposed by the market driven higher demand forces has led to an undue expansion of the education at this level with the larger role assumed by private sector popularly branded as self financing institutions. However, till date no effective regulatory mechanism has been initiated by any of the pro active government agencies and commissions to control the growth of these institutions who are prominently motivated by commercial interest but however successful in keeping in disguise their motives and interests. The quality of education at professional level is therefore, diluted undermining the prospects of employment of the students who are professionally qualified particularly the technically professional recognised under the ambit of engineers [1].This disproportionate rise in this particular area of education has resulted in an imbalance with poor quality of skilled manpower visible in the form of large army of unemployed youth.(Harikumar ,2009)[2]

It is also ascertained that the students are driven to professional courses lured by the prospects of employment which is abysmally very low in conventional courses. Hence the demand for the professional courses has disproportionately risen as also the increase in the number of institutions offering these courses. The lion share of this has been engrossed by the technical education. Therefore the impact of the initiation of the expansion of technical education and its consequent impact on the labour market especially with reference to the prospects of labour absorption and the incidence of unemployment is a matter of interest for any researcher.

The chronic problem of unemployment among engineers has strengthened the view of the people on the impotency of professional education as a guarantor of employment imposed due to various reasons and this observation holds relevant for both the diploma and graduate engineers in Kerala across various cross sections of the people hailing from the technical institutes (Sabu Thomas 2009)[3] . The present paper therefore aims to highlight the employment status of engineers in Kerala of 2009 with some focus on the higher dimensions of the problem of unemployment.

Technical Education, Scope and Importance: A Kerala Perspective

Technical education is that part of the stage of education offered at higher education level aiming at imparting technical skills to enhance the productivity of labour through innovations prospering mankind. The technical education imparted at various levels with the initial stage beginning from the technical schools is aimed at the creation of a skilled technical personnel and development of technical aptitude inculcating the spirit of science and technology for upgrading the technical base of the country capable of increasing production and economic growth. However the large chunk of population joining the technical education acquires technical knowledge only from the diploma level and presently it develops further at higher level if enrolment rate is accepted as the principle of judgment of the degree of preference revealed. The outnumbering of engineering colleges over the other institutions in recent times establishes the popularity of choice of education at undergraduate level for the technical courses particularly engineering diploma and degree courses and in India with wider scope of technical education being confined to engineering education the present study focuses on how far the engineering education in Kerala has succeeded in accomplishng the goals of ensuring maximum employment opportunity for both the engineering diploma holders and engineering graduates who are conferred a Diploma and B.Tech degree in the year 2009 in Kerala.

The technical educational in the state of Kerala comprising the diploma institutions and engineering colleges is spread across all the

14 districts in Kerala. The diploma engineering institutions outnumbered the engineering colleges in the beginning but the current strength of the engineering colleges is much higher than the diploma institutions as on the year 2009.The present study, therefore, is conducted from the basic information collected from the passed out diploma engineers and graduate engineers in Kerala having completed their respective courses during the year 2009 in the context of their employment status after two years of completion of the course on the basis of their relative strength.

The diploma institutions brought under the ambit of study consisted of 58 colleges in 27 disciplines irrespective of the characteristics of their type of management and the number of engineering colleges offering graduate level degree consisted of 94 colleges offering degree in 26 disciplines without characterizing students on the basis of their type of management.

The students passed out from diploma colleges in engineering in the year 2009 are 5702 and the students passed out from the graduate engineering colleges are 10991[4] and of the total passed out, at least 10 percentage of the passed out from each branch of study are selected as samples on random basis and further information is collected to know of their employment status after two years of passing out. Thus the sample strength at the diploma level consists of 664 numbers of students in 27 branches of study and at degree level there are 1237 number of students in 26 disciplines. The samples covers students randomly selected from all engineering diploma and degree colleges irrespective of any weighage assigned to the characteristic of students just other than their branch.

Employment

The Keynesian involuntary unemployment ideology scaffolds to distinguish between employment and unemployment where employment is considered to be a state where an able bodied individual is disposed with an opportunity of making their contribution towards production of goods and services resulting in the generation of income for their services rendered .The condition of unemployment is a state of involuntary idleness due to lack of work excluding idleness imposed due to labour disputes, vacations etc., caused by cyclical ,structural, neutral and other frictional factors. Here employment is also further classified as paid employee and self employed where paid employees are those not working in the capacity as employers but the self employed are those who are also working in the capacity of an employer.

The employment absorption of engineers for graduates and diploma holders in the labour market is abysmally low and many of the engineers are unemployed as the intake capacity of engineering institutions and its consequent outturns is disproportionately higher than the labour market demands or either the acquired skill is not in consonance with the labour market needs and the type of skill needed and skill supplied are different (Ramachandran 2002) [5]. The dynamism of market dependent employment reveals a wide disparity among students with respect to their employment status and it is found to be highly influenced by characteristics such as the branch of their study, their gender characteristics, the region of their study, etc[6] either at diploma level or at degree level and in some cases both at the diploma level and degree level. The employment status in the paper presents the current status of the students after two years of passing out in the form of accepting wage employment or otherwise called being in paid job , self employed, on contract job, seeking apprenticeship training , pursuing higher studies or either unemployed.. The present paper, therefore, aims to unveil some of the findings observed in this regard particularly the extent of employment and unemployment among engineers both at the diploma level and graduate level.

On Contract job

The contract job holders here refers to those group of engineers who are not permanently employed and hence are prevented from the benefits enjoyed under social security net provided to other permanent employees in the organization coupled with no guarantee of employment in the future. Hence engineers working on contract job cannot be treated as employed in its strict sense.

Apprenticeship Trainee

Many of the engineers undergo apprenticeship training to consolidate and strengthen their learning skills acquired through classroom learning by exposing themselves to practical based training in order to increase their scope of employment in their respective braches of study, arranged by the institution or through other governmental agencies.

Pursuing Higher studies

It is also observed that some of the students pursue higher studies rather than accepting employment either because of self interest or superimposed due to unemployment. A diploma engineer here may pursue Ꞵ.Tech degree and a B.Tech degree holder may pursue an M.Tech degree. The students pursuing other conventional courses after their respective engineering courses is also treated as students pursuing higher studies, for such courses also offer a degree higher to the level already acquired in their respective course of study.

Unemployed

The engineering students who fail to earn income devoid of potential employment source in their post study period are considered to be unemployed. The unemployed engineers here are involuntarily unemployed because after a waiting period of two years of completion of their study in respective courses, these students fail to get absorbed in the labour market. The incidence of unemployment is either due to no job offers or non acceptance of job from the non technical area with a paltry salary offer. The magnitude of the problem of unemployment among engineers is extensive in Kerala with high incidence of unemployment both among the diploma holders and graduate engineers. It is again observed that there exists a wide difference in the rate of unemployment with in each course, if the population is stratified into various categories with respect to their gender, region of their study, type of institution,etc. This paper emphasizes on the dimension of the problem of unemployment existing among engineers in Kerala as for the year 2009.

Methodology

The chisquare test of the independence is used in the present study to test the hypothesis that two categorical variables are independent of each other. The chisquare test is the popular method for testing hypotheses on discrete data that can be nominal or ordinal. The descriptive statistics like Arithmetic mean, Median...etc is meaningless here and thus the summary statistics useful for such data are frequencies and percentages. The chisquare statistics are used on such kind of data. A small chisquare statisitic indicates that the null hypothesis is correct and the two variables are independent of each other. Here we compare the observed cell frequencies with the expected cell frequencies. Observed cell frequencies are the actual number of cases falling in different cells of the contingency table and expected frequencies are the number of cases that should fall in each cell if there is no relationship between the two categorical variables. We reject the null hypothesis if 2-sided significance corresponding to the Pearson chi-square is less than the significance level selected (1%,5% or 10%). Thus a P value of less than .01 signifies that categorical variables are not related and their difference holds significant at 1% level. Similarly the P value of less than .05 holds significant at 5% level and the P value of less than 0.1 holds significant at 10% level.

Diploma and Graduate Engineers in Kerala of 2009 and their employment status on the branch of study

The diploma engineers and graduate engineers in Kerala for 2009 are studied with respect to their employment status and are analyzed across the various branches of their study. The diplomas engineers in 27 branches seek employment either as paid wages or are self employed or pursue higher studies either in technical or non technical stream of education. The paid job can be either permanent job or as apprentice training or on contract job. The rest of the engineers are unemployed. The status of these engineers is given in the table 1 below and discussed as per the results obtained.

The graduate engineers in 26 branches also seek employment as paid job with a marginal number of students as self employed. The other category of employment is either in the status of a contract job or apprenticeship trainee and the all the rest are unemployed with a major share of students looking for employment. The status of the graduate engineers on the observations made is recorded and its highlights are given in the table 2.

A close look at the figures presented in the table reveals that the incidence of unemployment is a fact both among the diploma holders and graduate engineers. This is substantiated by the relative distribution of the students across various occupations or forms of engagement in all branches of study for both the diploma holders and graduate engineers. A chisquare value of table I in the data unveils the employment status of the diploma engineers in Kerala of 2009 across various branches of their study. Among the diploma engineers it is observed that of the 664 number of students chosen as sample, 195 numbers of students are paid workers with none seeking self employment and the number of students who are unemployed engineers is 194 with 29.2 % of unemployment rate. Hence it is adequate to state that the problem of unemployment among diploma engineers is of higher intensity even if there is only relatively a marginal difference between the employed and unemployed in numbers. Similarly it is observed that the difference in terms of unemployment

Table 1 : Employment Status of the Diploma Engineers in Kerala of 2009 among Branches of study

	Branch of Study	Employed		On Contract Job		Apprenticeship/ Trainee Employee		Pursuing Higher Studies		Unemployed		Total	Chisquare value	P value
		No:	%	No:	%	No:	%	No:	%	No:	%			
1	Applied Electronics and Instrumentation	1	9.1	0	0	0	0	4	36.4	6	54.5	11	172.588[a]	.000*** Significant at 1% level
2	Architecture	2	40.0	0	0	0	0	1	20.0	2	40.0	5		
3	Automobile	10	43.5	0	0	0	0	7	30.4	6	26.1	23		
4	Bio Medical	4	50.0	0	0	0	0	2	25.0	2	25.0	8		
5	Computer Application	4	40.0	1	10.0	0	0	0	0	5	50.0	10		
6	Chemical	4	44.4	0	0	0	0	3	33.3	2	22.2	9		
7	Civil	35	47.9	3	4.1	0	0	20	27.4	15	20.5	73		
8	Commercial Practice	3	13.6	7	31.8	0	0	2	9.1	10	45.5	22		
9	Computer Science	30	33.3	4	4.4	4	4.4	31	34.4	21	23.3	90		
10	Computer Hardware Maintenance	7	18.4	4	10.5	3	7.9	11	28.9	13	34.2	38		
11	Electrical	18	25.0	10	13.9	5	6.9	24	33.3	15	20.8	72		
12	Electronics	21	21.0	5	5.0	8	8.0	31	31.0	35	35.0	100		
13	Electronics & Avionics	0		1	33.3	0	0	1	33.3	1	33.3	3		
14	Electronics & Communication	5	16.7	1	3.3	1	3.3	13	43.3	10	33.3	30		
15	Electronics & Instrumentation	2	11.8	2	11.8	1	5.9	7	41.2	5	29.4	17		

	Branch of Study	Employed		On Contract Job		Apprenticeship/ Trainee Employee		Pursuing Higher Studies		Unemployed		Total	Chisquare value	P value
		No:	%	*No:*	%	*No:*	%	*No:*	%	*No:*	%			
16	Electronics & Production Technology	3	30.0	0	0	1	10.0	3	30.0	3	30.0	10		
17	Information Technology	0	0	0	0	1	50.0	0	0	1	50.0	2		
18	Instrument Technology	1	16.7	0	0	1	16.7	1	16.7	3	50.0	6		
19	Mechanical	24	30.0	3	3.8	12	15.0	22	27.5	19	23.8	80		
20	Medical Electronics	4	36.4	0	0	0	0	3	27.3	4	36.4	11		
21	Polymer Technology	1	14.3	3	42.9	0	0	0	0	3	42.9	7		
22	Printing Technology	5	71.4	0	0	0	0	0	0	2	28.6	7	172.588[a]	.000*** Significant at 1% level
23	Quality Survey & Construction Management	2	100.0	0	0	0	0	0	0	0	0	2		
24	Telecommunication Technology	0	0	0	0	1	14.3	1	14.3	5	71.4	7		
25	Textile Technology	6	60.0	0		1	10.0	1	10.0	2	20.0	10		
26	Tool & Die Making	1	12.5	0		1	12.5	3	37.5	3	37.5	8		
27	Wood Technology	2	66.7	0		0	0	0		1	33.3	3		
	Total	195	29.4	44	6.6	40	6.0	191	28.8	194	29.2	664		

Source: *All tables and figures based on data collected by the investigator from the information obtained through field survey.*

among diploma holders across the various branches of study is significant at 1% level with a chi-square value of 172.588. The number of students pursuing higher studies is of 191 which is also of higher relevance showing that the difference between employed, unemployed and those pursuing higher studies has narrowed down possibly due to lack of employment opportunities or because students pursue higher studies to keep in disguise their unemployment status. Among the various branches of study, the number of unemployed is highest among the students of Electronics where as the unemployment rate is highest among the Applied Electronics and Instrumentation even if the number of unemployed in this particular branch is abysmally low with only 6 number of students unemployed. The lowest rate of unemployment is observable in the branch of Quality Survey and Construction Management with no unemployment as all students taken as random samples are found employed. The largest number of students pursuing higher studies as observed is in the branch of Computer Science and Electronics; where as the highest rate of enrolment for higher studies is from the branch of Electronic and Communication.. The rest of the students are either engaged in contract job or are undergoing Apprenticeship Training.

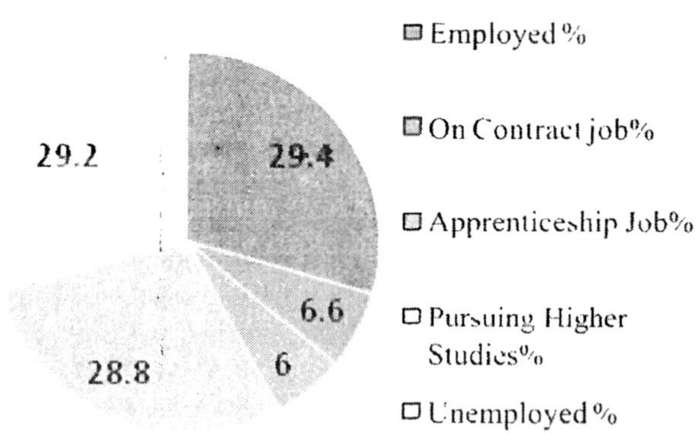

Figure 1: Employment Profile of Diploma Engineers (%) 2009

The overall employment Status of Diploma Engineers of 2009 as shown in Figure 1 reveals that of the total diploma engineers 29.4% are employed where as 29.2 % of engineers are unemployed, the diploma engineers are on contract job to the extent of 6.6%, with apprenticeship job of 6% and 28.8% of students pursuing higher studies.

Similarly the Figure 2 for Graduate Engineers depicts an employment rate of 54.6% with unemployment rate of 20.5% in total. The figure also reveals that the graduate engineers are seeking self employment to the extent of 0.3% and 5.3% of students are on contract job. The students seeking apprenticeship is of 1.1% and 18.1% of the students pursue higher studies.

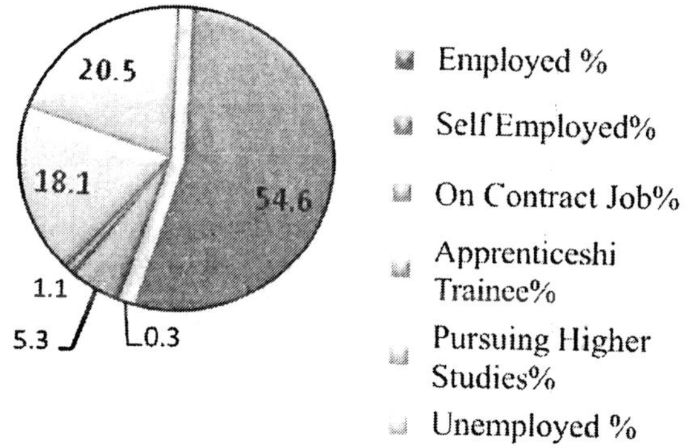

Figure 2 : Employment Profile of Graduate Engineers(2009)

The table 2 represents the data of the employment status for graduate engineers having completed B.Tech courses in the year 2009 in 26 branches of study. The table shows that among 1237 number of students taken as sample it is found that 676 numbers of students are in paid job with a marginal 4 number of students seeking self-employment. The number of students engaged in contract job are 65 and the number of students who are undergoing apprenticeship training is 14.The number of students pursuing higher studies is 224.The highest rate of unemployment among graduate engineers is

Table 2 : Employment Status of the Graduate Engineers in Kerala of 2009 among Branches of study

	Branch	EMPLOYEE STATUS														
		Employed		Self Employed		On Contract Job		Apprenticeship/ Trainee Employee		Pursuing Higher Studies		Unemployed		Total	Chi-square value	P value
		No:	%	No:	%	No:	%	No:	%	No:	%	No:	%	No:		
1	Agriculture Engineering	3	42.9	0		2	28.6	0		1	14.3	1	14.3	7		
2	Applied Electronics and Instrumentation	14	36.8	2	5.3	0		0		6	15.8	16	42.1	38		
3	Architecture	7	58.3	0		0		1	8.3	4	33.3	0		12		
4	Automobile	5	55.6	0		1	11.1	0		0		3	33.3	9		
5	Bio Medical	5	50.0	0		1	10.0	0		1	10.0	3	30.0	10		
6	Bio Technology	9	39.1	0		0		0		10	43.5	4	17.4	23		
7	Chemical	10	52.6	0		2	10.5	0		4	21.1	3	15.8	19		
8	Civil	46	57.5	1	1.2	3	3.8	1	1.2	[illegible]	11.2	20	25.0	80	249.616[a]	.000***(*** Significant at 1 % level)
9	Computer Science	127	54.7	0		7	3.0	0		39	16.8	59	25.4	232		
10	Electrical and Electronics	105	52.0	1	0.5	22	10.9	8	4.0	33	16.3	33	16.3	202		
11	Electronics and Communication	159	54.6	0		8	2.7	0		65	22.3	59	20.3	291		
12	Electronics and Instrumentation	5	35.7	0		2	14.3	2	14.3	3	21.4	2	14.3	14		
13	Information Technology	61	68.5	0		2	2.2	1	1.1	2	2.2	23	25.8	89		
14	Industrial	1	25.0	0		0		0		1	25.0	2	50.0	4		

	Branch	EMPLOYEE STATUS														
		Employed		Self Employed		On Contract Job		Apprenticeship/ Trainee Employee		Pursuing Higher Studies		Unemployed		Total	Chi-square value	P value
		No:	%	No:	%	No:	%	No:	%	No:	%	No:	%	No:		
15	Instrumentation	1	25.0	0		0		0		2	50.0	1	25.0	4	249.616a	.000***(*** Significant at 1 % level)
16	Instrumentation and Control	1	33.3	0		0		0		0		2	66.7	3		
17	Marine Engineering	2	50.0	0		2	50.0	0		0		0		4		
18	Mechanical	95	58.3	0		12	7.4	0		35	21.5	21	12.9	163		
19	Mechanical Automobile	2	66.7	0		0		0		1	33.3	0		3		
20	Mechanical Production	0		0		0		0		2	66.7	1	33.3	3		
21	Naval Architecture and Ship Building	2	66.7	0		0		0		1	33.3	0		3		
22	Polymer Engineering	2	50.0	0		0		1	25.0	1	25.0	0		4		
23	Polymer Science and Rubber Technology	2	66.7	0		0		0		1	33.3	0		3		
24	Printing	2	40.0	0		1	20.0	0		1	20.0	1	20.0	5		
25	Production Engineering	8	80.0	0		0		0		2	20.0	0		10		
26	Safety and Fire Engineering	2	100.0	0		0		0		0		0		2		
	Total	676	54.6	4	0.3	65	5.3	14	1.1	224	18.1	254	20.5	1237		

in the branch of Instrumentation and Control with many branches recording zero rate of unemployment such as Architecture, Marine engineering, Mechanical Automobile, Naval Architecture and Ship Building, Polymer engineering, Polymer Science and Rubber Technology, Production engineering and Safety & Fire engineering. The highest rate of employment absorption is in the branch of Safety and Fire engineering followed by the Production engineering and then Information Technology, Mechanical Automobile, Naval Architecture and Ship Building, Polymer Science and Rubber Technology.

District and Regional wise difference in the employment status of engineers among diploma holders and graduate engineers

In Kerala engineering education is spread all across 14 districts with different concentration of engineering institutions at diploma level and graduate level. The highest number of Diploma engineering institutes is in the district of Thrissur and the least number of engineering institutions in Wayanad district. For Graduate Engineering Colleges, the largest number of institutions is in the district of Thiruvananthapuram with again least number of colleges in the district of Wayanad[7].The students pursue their courses of study in various districts and the difference in the employment status among engineers across the various districts for both the diploma holders and graduate engineers of Kerala in the year 2009 is given in the table below.

The table 3 shows that for diploma engineers the highest percentage of unemployment is in Kollam district and the least unemployment rate is in the Wayanad district. The rate of unemployment in the districts of Thiruvananthapuram,Kottayam , Alappuzha ,Idukki and Malapuram is significantly above the state average of 29.2 %.The highest absorption rate with respect to employment is in Idukki district followed by Kasargode district. The contrasting observation of high employment and high unemployment in Idukki district is due to lower absorption in other aspects of employment taken such as in contract job, in apprenticeship trainee, in higher studies,etc. The difference in the employment status of engineers among diploma engineers is significant across district at 1% level with a chisquare value of 92.113.

A comparison of the employment rate and unemployment rate of engineers in the various districts of Kerala is given in the figure below for both diploma holders and graduate engineers. The figure 3 shows that for diploma engineers, the employment and unemployment rate is not uniform with some districts such as Thiruvananthapuram, Kollam, Alapuzha, Kottayam and Malapuram where the unemployment rate outnumber the employment rate and in all other districts the employment rate is higher than unemployment rate for the Diploma Engineers of Kerala in the year 2009. The difference in the employment and unemployment rate of diploma engineers shows that the location of the institution in certain respect might have influenced the employment prospect of engineers.

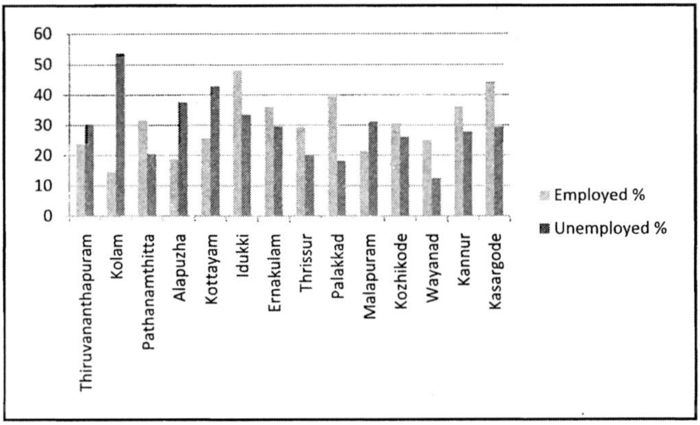

Figure 3 : District Wise Employment and unemployment rate of Diploma Engineers of 2009

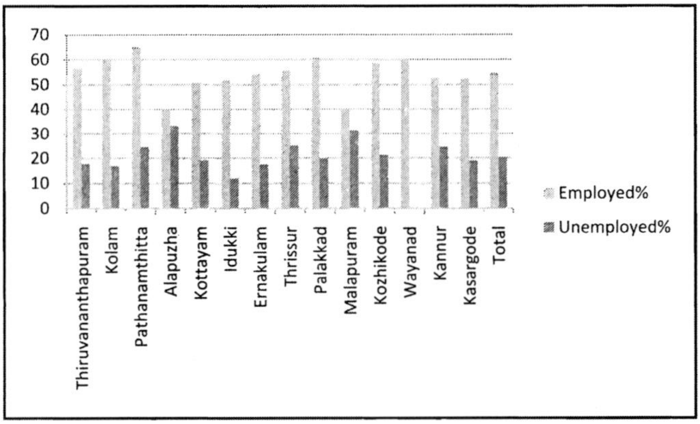

Figure 4 : District wise Employment and Unemployment rate of Graduate Engineers

Table 3 : Employment Status of the Diploma Engineers across the various districts in Kerala 2009

	District	Employed		On Contract Job		Apprenticeship/ Trainee Employee		Pursuing Higher Studies		Unemployed		Total	Chisquare value	P value
		No:	*%*	*No:*	*%*	*No:*	*%*	*No:*	*%*	*No:*	*%*			
1	Thiruvananthapuram	15	23.8	10	15.9	1	1.6	18	28.6	19	30.2	63	92.113[a]	.001***
2	Kolam	6	14.6	3	7.3	5	12.2	5	12.2	22	53.7	41		
3	Pathanamthitta	14	31.8	2	4.5	2	4.5	17	38.6	9	20.5	44		
4	Alapuzha	9	18.8	0		3	6.2	18	37.5	18	37.5	48		
5	Kottayam	9	25.7	2	5.7	3	8.6	6	17.1	15	42.9	35		
6	Idukki	13	48.1	1	3.7	2	7.4	2	7.4	9	33.3	27		
7	Ernakulam	22	36.1	3	4.9	4	6.6	14	23	18	29.5	61		
8	Thrissur	28	29.2	6	6.3	7	7.3	36	36	19	19.8	96		
9	Palakkad	13	39.4	4	12.1	4	12.1	6	18.2	6	18.2	33		
10	Malapuram	13	21.3	2	3.3	3	4.9	24	39.3	19	31.1	61		
11	Kozhikode	21	30.4	7	10.1	1	1.4	22	31.9	18	26.1	69		
12	Wayanad	4	25	3	18.8	1	6.2	6	37.5	2	12.5	16		
13	Kannur	13	36.1	1	2.8	4	11.1	8	22.2	10	27.8	36		
14	Kasargode	15	44.1	0		0		9	26.5	10	29.4	34		
	Total	195	29.4	44	6.6	40	6.0	191	28.8	194	29.2	664		

****.significant at 1% level*

Table 4 : Employment Status of the Graduate Engineers across the districts in Kerala 2009

	District	Employed		Self Employed		On Contract Job		Apprenticeship/ Trainee Employee		Pursuing Higher Studies		Unemployed		Total	Chi-square value	P value
		No:	%	No:	%	No:	%	No:	%	No:	%	No:	%	No:		
1	Thiruvananthapuram	102	56.7	1	0.6	1	1.7	0		42	23.3	32	17.8	180	83.666[a]	.059*
2	Kolam	60	60.0	0		0	5.0%	2	2.0	16	16.0	17	17.0	100		
3	Pathanamthitta	32	65.3	0		0	2.0	0		4	8.2	12	24.5	49		
4	Alapuzha	12	40.0	0		0	3.3	1	3.3	6	20.0	10	33.3	30		
5	Kottayam	53	51.0	0		0	7.7	0		23	22.1	20	19.2	104		
6	Idukki	26	52.0	0		0	4.0	3	6.0	13	26.0	6	12.0	50		
7	Ernakulam	143	54.0	3	1.1	3	6.0	4	1.5	52	19.6	47	17.7	265		
8	Thrissur	87	55.8	0		0	3.8	2	1.3	22	14.1	39	25.0	156		
9	Palakkad	37	60.7	0		0	9.8	0		6	9.8	12	19.7	61		
10	Malapuram	28	40.0	0		0	12.9	0		11	15.7	22	31.4	70		
11	Kozhikode	52	58.4	0		0	3.4	1	1.1	14	15.7	19	21.3	89		
12	Wayanad	3	60.0	0		0		0		2	40.0	0		5		
13	Kannur	30	52.6	0		0	7.0	0		9	15.8	14	24.6	57		
14	Kasargode	11	52.4	0		0	4.8	1	4.8	4	19.0	4	19.0	21		
	Total	676	54.66	4	0.3	65	5.3	14	1.1	224	18.1	254	20.5	1237		

**significant at 10% level*

Table 5 : Employment Status of the Diploma engineers in Kerala across regions

	REGION	EMPLOYMENT STATUS												
		Employed		On Contract Job		Apprenticeship/ Trainee Employee		Pursuing Higher Studies		Unemployed		Total	Chisquare value	P value
		No:	%	No:	%	No:	%	No:	%	No:	%			
1	Southern Kerala	53	22.9	17	7.4	14	6.1	64	27.7	83	35.9	231	12.175[a]	.144
2	Central Kerala	63	34.2	10	5.4	13	7.1	52	28.3	46	25.0	184		
3	Northern Kerala	79	31.7	17	6.8	13	5.2	75	30.1	65	26.1	249		
	Total	195	29.4	44	6.6	40	6.0	191	28.8	194	29.2	664		

Table 6 : Employment Status of the Graduate Engineers of Kerala 2009 across the regions

	Region	Employed		Self Employed		On Contract Job		Apprenticeship/ Trainee Employee		Pursuing Higher Studies		Unemployed		Total	Chi-square value	P value
		No:	%	No:	%	No:	%	No:	%	No:	%	No:	%	No:		
1	Southern Kerala	259	55.9	1	0.2	18	3.9	3	0.6	91	19.7	91	19.7	463	15.471[a]	.116
2	Central Kerala	256	54.4	3	0.6	24	5.1	9	1.9	87	18.5	92	19.5	471		
3	Northern Kerala	161	53.1	0		23	7.6	2	0.7	46	15.2	71	23.4	303		
	Total	676	54.666666	4	0.3	65	5.3	14	1.1	224	18.1	254	20.5	1237		

Table 7 : Employment status of the male and female Diploma engineers of 2009 in Kerala

	SEX	EMPLOYMENT STATUS												
		Employed		On Contract Job		Apprenticeship/ Trainee Employee (%)		Pursuing Higher Studies (%)		Unemployed		Total	Chisquare value	P value
		No:	%	No:	%	No:	%	No:	%	No:	%			
1	Male	136	32.5	21	5	32	7.7	130	31.1	99	23.7	418	27.174[a]	.000***
2	Female	59	24	23	9.3	8	3.3	61	24.8	95	38.6	246		
	Total	195	29.4	44	6.6	40	6	191	28.2	194	29.2	664		

****significant at 1% level*

The figure 4 shows the district wise employment and unemployment rate of graduate engineers in Kerala during the year 2009, but unlike in the case of diploma engineers, for the graduate engineers the rates of employment is relatively higher than the rate of unemployment across all the districts in Kerala.

The employment status of the graduate engineers of Kerala of 2009 is given in Table 4. The table reveals that the difference in employment status of graduate engineers in Kerala is significant at 10% level. The table also reveals that the highest rate of unemployment is in the district of Alapuzha with 33.3 % and zero rate of employment in the district of Wayanad and the highest rate of employment absorption in Pathanamthitta district with 65.3% and a relatively lower rate of employment absorption in Alapuzha and Malapuram district with employment absorption of only 40% for graduate engineers for both these districts.

Regional Wise Employment Status of Diploma and Graduate engineers in Kerala

The regional distribution of the employment status of the diploma engineers and graduate engineers is shown to know of any difference in the employment status among students classified on the basis of regions of their study. Southern Kerala consists of the districts of Thiruvananthapuram. Kollam, Pathanamthitta, Alapuzha and Kottayam while Central Kerala consists of Idukki, Ernakulam and Thrissur districts while the Northern Kerala consists of Malapuram, Kozhikode, Wayand, Kannur and Kasargode districts.

The table 5 reveals that for diploma engineers the highest rate of unemployment is in Southern Kerala followed by Northern Kerala and the least unemployment is observable in Central Kerala. The highest employment absorption is also observed in Central Kerala with 34.2 % followed by Northern Kerala and Southern Kerala. But the difference in the employment status of the diploma holders is however not significant even at 10% level across the regions in Kerala

The difference in the employment status of the graduate engineers is shown in Table 6 which shows that the highest rate of unemployment is in the Northern region of Kerala with 23.4 % followed by Southern and Central Kerala. The highest employment absorption is in Southern Kerala with 55.9% followed by Central and Northern Kerala with the difference in the employment status of the engineers not being found significant even at 10% level.

The table 6 also reveals that there is not much difference in the number of students pursuing higher studies across the various regions of Kerala.

Gender difference in the employment status among diploma holders and graduate engineers

The engineering profession was gender biased in earlier times with a conventional belief that engineering education is favourable only to men characterised by low women enrolment in these area of study, but as time passed, it is observed that women folks do not stay away from the engineering profession, instead there is an observable and significant entry of females into this profession with higher enrolment ratio. The higher enrolment ratio however does not ensure that the labour market is unbiased towards female engineers if viewed from the perspective of employment. The table 7 and 8 highlight on the employment status of the engineers based on their gender characteristics for both the diploma and graduate engineers.

The table 7 shows that among the diploma engineers females are relatively more unemployed than males and the employment absorption of males is also higher than the females and it is also observed that the employment status of the diploma engineers across gender is significantly different at 1% level with a chi-square value of 27.174 and it is also found that the unemployment rate of female diploma holders is to the extent of 38.6% high above the state wise unemployment rate of 29.2% among the diploma holders.

Table 8 given below presents the employment status of graduate engineers of 2009. Here also the unemployment rate of females is higher than the male engineers with 26.8% and 14.8% respectively.

The employment absorption in the labour market reveals that 61% of males are employed as paid workers, 0.3% self employed and female engineers are absorbed to the level of 47.7%. It is also observed that males are on contract job to the extent of 6.1% and female engineers are on contract job to the extent of 4.4 %.But it is interesting to note that rate of female engineers pursuing higher studies is 20.2 % with male engineers pursuing higher studies to the extent of 16.1%.It is also observed that the employment status of graduate engineers among male and females significantly differ at 1% level with a chisquare value of 40.306.

Table 8 : Employment status of the male and female Graduate Engineers of 2009 in Kerala

SEX		EMPLOYMENT STATUS														
		Employed		Self Employed		On Contract Job		Apprenticeship/ Trainee Employee		Pursuing Higher Studies		Unemployed		Total	Chi-square value	P value
		No:	%	No:	%	No:	%	No:	%	No:	%	No:	%	No:		
1	Male	393	61.0	2	0.3	39	6.1	11	1.7	104	16.1	95	14.8	644	40.306ª	.000***
2	Female	283	47.7	2	0.3	26	4.4	3	0.5	120	20.2	159	26.8	593		
Total		676	54.6	4	0.3	65	5.3	14	1.1	224	18.1	254	20.5	1237		

****significant at 1% level*

Table 9: Employment Status of the Diploma Engineers among the different social categories

	SOCIAL CATEGORY	EMPLOYMENT STATUS												
		Employed		On Contract Job		Apprenticeship/ Trainee Employee (%)		Pursuing Higher Studies (%)		Unemployed		Total	Chisquare value	P value
		No:	%	No:	%	No:	%	No:	%	No:	%			
1	General	50	25.6	6	3.1	12	6.2	75	38.5	52	26.7	195	27.031[a]	.001*** Signi ficant at 1 % level
2	Other Backward Class	134	32.4	33	8	20	4.8	100	24.2	127	30.7	414		
3	SC & ST	11	20	5	9.1	8	14.5	16	29.1	15	27.3	55		
	Total	195	29.4	44	6.6	40	6	191	28.2	194	29.2	664		

Social category and Employment status of Engineers

The social fabric of Kerala like other parts of the country is also characterized by the existence of various castes and sub castes belonging to different communities. The most popular classification of the Indian society is such that the society is divided into Forward castes called the General category, Scheduled Castes and Scheduled Tribes and Other Backward classes. However, the classification of the society on the basis of their educational backwardness is not explicitly observable in terms of number of people enrolling for higher studies in Kerala except for the scheduled tribes but that same may not hold true with respect to their absorption in the labour market. It is therefore interesting to unveil the influence of social milieu of an individual in affecting his employment profile. The Table 9 reveals the employment status of diploma engineers of Kerala across the various social categories.

The table 9 shows that the highest rate of unemployment is among the backward class particularly the students belonging to OBC category with 30.7% rate of unemployment followed by SC and ST students with 27.3%. However it is a paradox to observe that the highest employment absorption is also from the students belonging to OBC category. The difference in employment status of the diploma students is significant at 1% level with a chi square value of 27.031 .The table 10 shows the employment status of the Graduate engineers of Kerala among the different social categories.

The table 10 shows that the unemployment rate is highest among the students belonging to other backward classes with 25.4% again followed by the SC and ST students with 23.1% where as the highest employment absorption is among the students belonging to the General category. The difference in the employment status of the students is significant at 1 % level with a chisquare value of 34.956.

Institutional characteristics and Employment status of Engineers

Engineering education in Kerala is imparted through three forms of institutions, one owned directly by the government, the other through the aided institutions and last through the self financing institutions and these three type of institutions exist both for the diploma institutions and B.Tech institutions. In the initial stages of the development of these technical institutions, the government institutions outnumbered the private institutions, for the policy of government did not favour much private participation in order to have a complete control on the higher education particularly in the technical education even if the profile changes during the post reform period with more number of private unaided institutions appearing in the field of technical education. As for the students of 2009 passed out of the diploma institutions taken from 49 government colleges, 6 Aided colleges and 3 Private unaided colleges and that of engineering Graduate colleges taken from 38 Government colleges including University departments, 3 Aided colleges and 53 Private unaided colleges, their employment status for the diploma students, classified on institutional characteristics is given below in table 11.

The table shows that the rate of unemployment is marginally higher for students in Government colleges followed by the Private unaided colleges and Private aided colleges. The highest employment absorption is seen among the students from private unaided colleges. The table also shows that the employment status of the diploma engineers of 2009 classified on their type of institution of study significantly differ at 5% level with a chi square value of 17.749

Table 12 shows the employment status of Graduate engineers of 2009 across the students hailing from various types of Institutions. The table shows that the highest rate of unemployment is among the students from the Private unaided colleges followed by the students from the Government colleges and the least incidence of unemployment is from among the students of the Private aided colleges.

The table also shows that the highest rate of employment absorption is also from the Private aided Colleges and the least employment absorption is again from the Private unaided colleges. The difference in the employment status among the graduate engineers of 2009 with a chisquare value of 23.118 is also significant at 1% level.

Table 10 : Employment Status of the Graduate Engineers among the different social categories

	SOCIAL CATEGORY	EMPLOYMENT STATUS															
		Employed		Self Employed		On Contract Job		Apprenticeship/ Trainee Employee		Pursuing Higher Studies		Unemployed		Total	Chi-square value	P value
		No:	%	No:	%	No:	%	No:	%	No:	%	No:	%	No:		
1	General	367	58.8	3	0.5	20	3.2	7	1.1	127	20.4	100	16.0	624	34.956[a]	.000***
2	Other Backward Class	281	51.3	1	0.2	38	6.9	6	1.1	83	15.1	139	25.4	548		
3	SC & ST	28	43.1	0		7	10.8	1	1.5	14	21.5	15	23.1	65		
	Total	676	54.6	4	0.3	65	5.3	14	1.1	224	18.1	254	20.5	1237		

****significant at 1% level*

Table 11 : Employment status of the Diploma engineers in Kerala across the type of Institutions

	TYPE OF INSTITUTION	EMPLOYMENT STATUS												
		Employed		On Contract Job		Apprenticeship/ Trainee Employee		Pursuing Higher Studies		Unemployed		Total	Chisquare value	P value
		No:	%	No:	%	No:	%	No:	%	No:	%			
1	Govt College	156	30.1	41	7.9	33	6.4	134	25.9	154	29.7	518	17.749[a]	.023**
2	Pvt Aided College	28	23.9	2	1.7	7	6.0	48	41.0	32	27.4	117		
3	Pvt Unaided College	11	37.9	1	3.4	0	0	9	31.0	8	27.6	29		
	Total	195	29.4	44	6.6	40	6.0	191	28.8	194	29.2	664		

***significant at 5%*

Table 12 : Employment status of the Graduate engineers in Kerala across the type of Institutions

	TYPE OF INSTITUTION	EMPLOYMENT STATUS														
		Employed		Self Employed		On Contract Job		Apprenticeship/ Trainee Employee		Pursuing Higher Studies		Unemployed		Total	Chi-square value	P value
		No:	%	No:	%	No:	%	No:	%	No:	%	No:	%	No:		
1	Govt College/IHRD	315	57.3	2	0.4	27	4.9	7	1.3	105	19.1	94	17.1	550	23.118[a]	.010***
2	Private Aided College	74	62.2	0		6	5.0	3	2.5	22	18.5	14	11.8	119		
3	Private Unaided College	287	50.5	2	0.4	32	5.6	4	0.7	97	17.1	146	25.7	568		
	Total	676	54.6	4	0.3	65	5.3	14	1.1	224	18.1	254	20.5	1237		

****significant at 1% level*

Findings and Conclusion

The present study therefore observes that the incidence of educated unemployment which was earlier limited only to the conventional courses has also gone to the extent of being evident in higher levels of technical education. The study highlights on some of the major findings.

The problem of unemployment among engineers is acute for both the diploma holders and graduate engineers where the rate of unemployment of 29.2% among the diploma holders is greater than the graduate engineers unemployment rate of 20.5% for 2009.While classifying the population on the basis of the branches of study , the sample study observes that the difference in the employment status among engineers for both the diploma holders and graduate engineers is significant at 1% level thereby substantiating the fact that unemployment is affected by the course of study chosen manifested due to high mismatch in demand and supply factors of engineering education in various branches. Apart from branch the other important factors that influence the employment status of students is the location of institution classified on the basis of districts particularly for the diploma holders even if such do not hold true for students classified on the basis of the region of their study. For diploma engineers the highest unemployment is in Kollam district with 53.7% rate of unemployment and for graduate engineers, the highest unemployment rate is in Alapuzha district with 33.3 % and the difference in the employment status among engineers for both the diploma holders and graduate engineers is significant at 1% level and 10% level respectively. The categorization of the population according to the regions reveals that for the diploma engineers, highest unemployment rate is in Southern Kerala with 35.9% and for the graduate engineers, the highest unemployment rate is in Northern Kerala with 23.4% but the difference in the employment status of engineers classified on the basis of the regions for both the diploma and graduate engineers is not significant even at 10% level.

It is also observed that the gender characteristics of the students also influence the prospects of employment and hence the incidence of unemployment is found more among females with unemployment rate of 38.6% for diploma engineers and 26.8% for graduate engineers where as the male unemployment rate is 23.7% for diploma engineers and 14.8% for graduate engineers with significant difference in employment status both for the diploma holders and graduate engineers at 1% level. The other important factor that influences the incidence of unemployment is with respect to the social category of students where unemployment is higher among Other Backward class students for both diploma engineers and graduate engineers with 30.7% and 25.4% rate of unemployment respectively. The rate of unemployment among the general category students is 26.7% for diploma engineers and for the graduate engineers the unemployment rate is 16%. The rate of unemployment among the SC&ST diploma engineers is 27.5% and for graduate engineers belonging to the same category the unemployment rate is 23.1 % and the difference in the employment status among students classified on the basis of the various social category is significant at 1% for both the diploma engineers and graduate engineers.

At last the observable difference in the type of institution also do affect the employment status and the unemployment rate is higher among the students from the government colleges for diploma holders with 29.7% and for graduate degree holders unemployment rate is higher among the students from private unaided colleges with 25.7%. At diploma level the unemployment rate among the students coming from private aided college students is 27.4% and for private unaided colleges the rate of unemployment is 27.6% For graduate engineers the unemployment rate among the Government college students is 17.1% and for students among private aided college the rate of unemployment is 11.8% and difference in employment status among student classified on the basis of the institution of their study is significant at 1% level for both the diploma and graduate engineers.

References

1. Anil K Yadav, V K Berry and Jerry Joseph(2002), The Indian Journal of Labour Economics, Vol 45, No4,2002
2. Athreye, Suma S (2005): Human Capital, Labour Scarcity and Development of the Software Services Sector' in A Saith and M Vijauabaskar (eds)ICTs and Indian Economic Development: Economy, Work, regulation, Sage Publications, New Delhi,pp154-174
3. Ashok Mathur and Rajendra P Mamgain, Technical Skills, Education and Economic Development in India , The Indian Journal of Labour Economics, Vol45,No4,2002

4. B S Prakash and K P Gupta, Employability Pattern of ITI Graduates: A Profile of the Vocational Education System, The Indian Journal of Labour Economics, Vol.45, No.4,2002

5. Carol Upadhya, Employment, Exclusion and Merit in the Indian IT Industry, Economic and Political Weekly, May 19,2007

6. CJ Fuller, Haripriya Narasimhan (2006) Engineering Colleges,'Exposure and Information Technology, Professionals in Tamilnadu, Economic and Political Weekly 2006.

7. Deshpande ,S and Yadav (2006), Redesigning Affirmative Action: Castes and Benefits in Higher Education, Economic and Political Weekly, Vol41(1)pp121-50

8. Harikumar S & Ramadevi T , Higher Education System in Kerala An Overview,in' Technical Education in Kerala Emerging Trends, Serials Publications, New Delhi,2009

9. Hone Angus(1968),Unemployed Engineers, Economic and Political Weekly, April 13,1968

10. Hone Angus(1968), What kind of Manpower Planning ; Economic and Political Weekly, July 27,1968

11. Jeemol Unni and Uma Rani, Technical Change and Workforce in India(2004): Skill Based Growth,The Indian Journal of Labour Economics, Vol47, No.4,2004

12. Padmanabhan, Nirmala (2008), software industry and Gender Inclusive Development under Globalisation- A Perspective from Kerala.

13. Ramachandran,H,(2002), Education Skill Development and changing labour market , The Indian Journal of Labour Economics,Vol 45 No 4, 2002

14. Sabu Thoms and M K Shelly, Engineering Labour Market: Employment Scenario of the Fresh Engineers in the Pre and Post Economic Recession Period- An Analysis of Kerala, Productivity, Vol.50, Oct,Dec 2009

15. Sajitha Bashir(2002), Upgrading The Educational Attainment and Skill Level of Karnataka Work force: Challenges and Options, The Indian Journal of Labour Economics, Vol.45, No.4,2002

16. Shashi Bhala(2004), Skill Development System In India, The Indian Journal of Labour Economics, Vol 47,No.4,2004

Endnotes

1. Hone Angus, Economic and Political Weekly, April 13, 1968
2. Harikumar S & Ramadevi T , Higher Education System in Kerala An Overview,in Technical Education in Kerala Emerging Trends, Serials Publications, New Delhi, 2009
3. Sabu Thoms and M K Shelly, Engineering Labour Market: Employment Scenario of the Fresh Engineers in the Pre and Post Economic Recession Period- An Analysis of Kerala, Productivity, Vol.50, Oct,Dec 2009
4. Data compiled by the investigator from the information obtained from engineering colleges appeared for examination in the year 2009 as regular students.
5. H Ramachandran, Education Skill Development and changing labour market , The Indian Journal of Labour Economics,Vol 45 No 4, 2002
6. Ashok Mathur and Rajendra P Mamgain, Technical Skills, Education and Economic Development in India , The Indian Journal of Labour Economics, Vol 45, No4,2002
7. Data compiled from the Annual Technical Manpower Review 2008, NTMIS,NODAL Centre,CUSAT

❑❑❑

Index

E

F

G

H

I

Q

R

S

T